CLIMATE CHANGE AND ITS CAUSES, EFFECTS AND PREDICTION

CLIMATE CHANGE LITIGATION AND LAW

Climate Change and its Causes, Effects and Prediction

Global Climate Change
Horace M. Karling (Editor)
2001. ISBN: 1-56072-999-6

Global Climate Change Revisited
Harace B. Karling (Editor)
2007. ISBN: 1-59454-039-X

Climate Change Research Progress
Lawrence N. Peretz (Editor)
2008. ISBN: 1-60021-998-5

Climate Change: Financial Risks
United States Government Accountability Office
2008. ISBN: 978-1-60456-488-4

Post-Kyoto: Designing the Next International Climate Change Protocol
Matthew Clarke
2008. ISBN: 978-1-60456-840-0
2008. ISBN: 978-1-61668-101-2 (E-book)

The Effects of Climate Change on Agriculture, Land Resources, Biodiversity in the United States
Peter Backlund, Anthony Janeto and David Schimel
2009. ISBN: 978-1-60456-989-6

Land Use and Climate Change
Suresh C. Rai
2009. ISBN: 978-1-60741-362-2

Economics of Policy Options to Address Climate Change
Gregory N. Bartos
2009. ISBN: 978-1-60692-116-6

Thresholds of Climate Change in Ecosystems
Vicente Orostegui (Editor)
2009. ISBN: 978-1-60741-487-2

Climate Variability, Modeling Tools and Agricultural Decision-Making
Angel Utset (Editor)
2009. ISBN: 978-1-60692-703-8
2009. ISBN: 978-1-60876-791-5 (E-book)

Carbon Offsets: Examining their Role in Greenhouse Gas Reduction
Karen T. Morningstar (Editor)
2010. ISBN: 978-1-60741-444-5

Emissions Trading: Lessons Learned from the European Union and Kyoto Protocol Climate Change Programs
Ervin Nagy and Gisella Varga (Editors)
2009. ISBN: 978-1-60741-194-9

Constructing Climate Change Legislation: Background and Issues
Gerald P. Overhauser (Editor)
2009. ISBN: 978-1-60692-986-5

Designing Greenhouse Gas Reduction and Regulatory Systems
Sonja Enden (Editor)
2009. ISBM: 978-1-60741-195-6

Global Climate Change: International Perspectives and Responses
Elias D'Angelo (Editor)
2009. ISBN: 978-1-60741-233-5

Disputing Global Warming
Anton Horvath and Boris Molnar (Editors)
2009. ISBN: 978-1-60741-235-9
2009. ISBN: 978-1-60876-503-4 (E-book)

Focus on Climate Change and Health
Viroj Wiwanitkit
2009. ISBN: 978-1-60741-247-2

Impacts of Climate Change on Transportation and Infrastructure A Gulf Coast Study
Iason Pavlopoulos (Editor)
2009. ISBN: 978-1-60741-424-7

Coastal Sensitivity to Sea Level Rise - Focusing on the Mid-Atlantic Region
Melvin C. Urajner (Editor)
2009. ISBN: 978-1-60741-440-7

Policy Option Issues for CO2 Emissions
Nikolaus Vogler (Editor)
2010. ISBN: 978-1-60741-381-3

Lightning in the Tropics: From a Source of Fire to a Monitoring System of Climatic Changes
Osmar Pinto, Jr.
2009. ISBN: 978-1-60741-764-4

Carbon Capture and Greenhouse Gases
Imrus Juhász and Gyorgy Halász (Editors)
2010. ISBN: 978-1-60692-089-3

The Science of Climate Change and Policy Implications
Kyle S. Hartzell (Editor)
2010. ISBN: 978-1-60741-448-3

Effects of Climate Change on Energy Production and Use in the U.S.
Alrik M. Solberg (Editor)
2010. ISBN: 978-1-60741-426-1

The Role of Agriculture in Carbon Capture and Climate Change
Arvid Bjurstrom (Editor)
2010. ISBN: 978-1-60741-445-2

Carbon Sequestration: Methods, Modeling and Impacts
Elke Hoch and Siegbert Grunwald (Editors)
2010. ISBN: 978-1-60741-498-8

Climate Change Litigation and Law
Jean-François Masson (Editor)
2010. ISBN: 978-1-60876-089-3

Carbon Tax and Cap-and-Trade Tools: Market-Based Approaches for Controlling Greenhouse Gases
Nelson E. Burney (Editor)
2010. ISBN: 978-1-60876-137-1

Global Change and Forestry: Economic and Policy Impacts and Responses
Jianbang Gan, Stephen Grado and Ian A. Munn (Editors)
2010. ISBN: 978-1-60876-262-0

Assessing Climate Change Impacts on the United States
Alex B. McNeill (Editor)
2010. ISBN: 978-1-60876-160-9
2010. ISBN: 978-1-60876-160-9 (E-book)

Climate Change Measures and Trade Considerations
George R. Fried (Editor)
2010. ISBN: 978-1-60876-756-4

Responding to Impacts of Climate Change on Water Resources
Zachary E. Quinn (Editor)
2010. ISBN: 978-1-60741-992-1
2010. ISBN: 978-1-61728-095-5 (E-book)

Climate Change: Fundamental Issues and Policy Tools
Elise M. Farrugia (Editor)
2010. ISBN: 978-1-60741-997-6
2010. ISBN: 978-1-61728-052-8 (E-book)

Effects of Climate Change on Aquatic Invasive Species
Sofia A. Contreras (Editor)
2010. ISBN: 978-1-61728-005-4
2010. ISBN: 978-1-61728-251-5 (E-book)

CLIMATE CHANGE AND ITS CAUSES, EFFECTS AND PREDICTION

CLIMATE CHANGE LITIGATION AND LAW

JEAN-FRANÇOIS MASSON
EDITOR

Nova Science Publishers, Inc.
New York

For permission to use material from this book please contact us:
Telephone 631-231-7269; Fax 631-231-8175
Web Site: http://www.novapublishers.com

LIBRARY OF CONGRESS CATALOGING-IN-PUBLICATION DATA

Climate change litigation and law / editor, Jean-François Masson.
p. cm.
Includes index.
ISBN 978-1-60876-089-3 (hardcover)
1. Climatic changes--Law and legislation--United States. I. Masson, Jean-François.
KF3775.C55 2009
344.7304'6342--dc22

2009041956

Published by Nova Science Publishers, Inc. † New York

CONTENTS

PREFACE

Climate change is viewed as a global issue, but proposed responses require action at the national level. Over the past 16 years, a variety of voluntary and regulatory actions have been proposed or undertaken in the United States, including monitoring of electric utility carbon dioxide emissions, improved appliances efficiency, and incentives for developing renewable energy sources. The Energy Policy Act of 2005 included provisions indirectly related to greenhouse gas emissions, such as energy efficiency and renewable energy. The Energy Independence and Security Act of 2007 addresses renewable energy and conservation, but also includes provisions specifically on climate change. This book emphasizes the steps that need to be taken to reduce the growth of greenhouse gas emissions. The authors of this book also discuss international law aspects of a nation's contributions to climate change, and the few international claims made against the United States to date.

This book consists of public documents which have been located, gathered, combined, reformatted, and enhanced with a subject index, selectively edited and bound to provide easy access.

Chapter 1 - The scientific, economic, and political questions surrounding climate change have long been with us. This report focuses instead on a relative newcomer: the legal debate. Though the first court decision related to climate change appeared 18 years ago, such litigation has proliferated in just the past five.

The court cases, decided and pending, arise in seven contexts. The first is the Clean Air Act (CAA). In April, 2007, the Supreme Court held in *Massachusetts v. EPA* that at least as to mobile sources of emissions (cars, trucks), EPA has authority under the CAA to regulate greenhouse gas (GHG) emissions. This decision puts pressure on EPA to move forward as well with regulation of GHGs from stationary sources (powerplants, factories).

Second, litigation under the Marine Mammal Protection Act, and current and likely litigation under the Endangered Species Act, raise the possibility that the impacts of climate change on wildlife may constrain government and private activities associated with GHG emissions.

Third, what about federal energy statutes? A much-publicized decision held that under the Energy Policy and Conservation Act, the United States must monetize the benefits of reduced carbon emissions as part of setting light-truck fuel economy standards. The Outer Continental Shelf Lands Act also has been invoked.

Fourth, various statutes requiring federal government analysis and information dissemination — the National Environmental Policy Act (NEPA), Global Change Research Act, and Freedom of Information Act — have generated climate-change litigation. NEPA suits make up the most numerous category of climate-change litigation. Courts appear to agree that if a plaintiff can establish standing, NEPA can be used to compel agency consideration of the climate change effects of its actions.

Fifth, can common law tort theories such as nuisance be used to force cutbacks in GHG emissions, or payment of damages? All four decisions thus far have denied relief to plaintiffs on lack of standing or political question grounds.

Sixth are the suits challenging state regulation of GHG emissions from motor vehicles as preempted by the federal corporate average fuel economy standards or federal authority over foreign policy. The two rulings thus far have rejected these challenges. Moreover, these issues will be moot unless EPA's denial of California's request for a waiver of federal preemption under the CAA is overturned in court.

Seventh, chiefly with respect to coal-fired power plants, what role will be played by state utilities laws?

Finally, the report discusses international law aspects of a nation's contributions to climate change, and the few international claims made against the United States to date.

Chapter 2 - California has adopted regulations requiring new motor vehicles to reduce emissions of greenhouse gases (GHGs), beginning in model year 2009. The Clean Air Act (CAA) generally preempts states from adopting their own emission standards for mobile sources. However, the act allows such standards in California, if the state obtains a waiver of CAA preemption from EPA.

California requested this waiver in 2005, but EPA took until December 19, 2007, to decide that it would deny the request. On that day, EPA Administrator Stephen Johnson wrote California Governor Schwarzenegger to say, "I have decided that EPA will be denying the waiver and have instructed my staff to draft appropriate documents setting forth the rationale for this denial in further detail.... " According to press reports, the decision was taken against the unanimous advice of the agency's technical and legal staffs. The Administrator published a decision document denying the waiver in the March 6, 2008 *Federal Register*.

Following EPA's denial of the waiver request, California and environmental groups petitioned for review in the D.C. Circuit, with 18 other states intervening on California's side. The interest of the intervening states derives from the fact that under the CAA, other states may adopt motor vehicle emission standards identical to California's and avoid CAA preemption if California is granted a waiver. Fourteen states and the District of Columbia have already adopted such regulations.

This report reviews the nature of EPA's, California's, and other states' authority to regulate emissions from mobile sources, the applicability of that authority to GHGs, and issues related to the California waiver request. To obtain a waiver, California must meet conditions laid out in CAA Section 209(b): the state must first determine that its standards will in the aggregate be at least as protective of public health and welfare as applicable federal standards. The EPA Administrator must then find whether the state's determination is arbitrary and capricious; whether the state needs the standards to meet compelling and extraordinary conditions; and whether the standards and accompanying enforcement procedures are consistent with CAA Section 202(a).

This report does not analyze whether California is preempted from regulating mobile-source GHGs by the Corporate Average Fuel Economy (CAFE) requirements of the Energy Policy and Conservation Act of 1975, or the amended CAFE standards of the Energy Independence and Security Act of 2007 (P.L. 110-140). Under these laws, authority to set fuel economy standards is reserved to the federal government—specifically, the National Highway Traffic Safety Administration (NHTSA). In several court cases and in other venues, the auto industry is maintaining that the regulation of mobile-source GHG emissions is simply another method of regulating fuel economy, so California's GHG standards (and identical standards adopted by other states) are preempted. Two federal district courts have rejected this argument, but both decisions have been appealed.

On January 21, 2009, California submitted a formal request to President Obama and EPA Administrator-designate Lisa Jackson for reconsideration of the waiver denial. In response to this request, EPA will hold a hearing March 5, 2009, and will accept written comments until April 6.

Chapter 3 - Climate change is viewed as a global issue, but proposed responses generally require action at the national level. In 1992, the United States ratified the United Nations Framework Convention on Climate Change (UNFCCC), which called on industrialized countries to take the lead in reducing greenhouse gases. Over the past 16 years, a variety of voluntary and regulatory actions have been proposed or undertaken in the United States, including monitoring of electric utility carbon dioxide emissions, improved appliance efficiency, and incentives for developing renewable energy sources. This report provides background on the evolution of U.S. climate change policy, from ratification of the UNFCCC to the George W. Bush Administration's 2001 rejection of the Kyoto Protocol to the present. Recent federal court decisions—most notably the Supreme Court's 2007 decision in *Massachusetts v. EPA* that the Environmental Protection Agency has the authority to regulate motor vehicle greenhouse gas emissions under the Clean Air Act—have raised the issue of whether EPA should directly regulate greenhouse gases. This report focuses on major regulatory programs that monitor or reduce greenhouse gas emissions, along with their estimated effect on emissions levels.

The George H. W. Bush, Clinton, and George W. Bush Administrations largely relied on voluntary initiatives to reduce the growth of greenhouse gas emissions. This focus was particularly evident in the current Administration's 2006 Climate Action Report (CAR), submitted under the provisions of the UNFCCC. Of roughly 50 programs summarized in the 2006 CAR, seven were described as "regulatory." However, this small subset of the total U.S. effort accounts for a large share of greenhouse gas emission reductions achieved over the past decade-and-a-half. In general, these efforts were established and implemented in response to concerns other than climate change, such as energy efficiency and air quality.

The Energy Policy Act of 2005 (P.L. 109-58) included provisions indirectly related to greenhouse gas emissions, such as energy efficiency and renewable energy. The Energy Independence and Security Act of 2007 (P.L. 110-140) addresses renewable energy and conservation, but also includes provisions specifically on climate change. These include a requirement for the use of renewable fuels with lower lifecycle greenhouse gas emissions than petroleum fuels, and the establishment of an Office of Climate Change and Environment in the Department of Transportation to implement research on mitigating the causes and addressing the effects of climate change on transportation. In June 2008, the Senate considered a bill (S. 3036) to establish an economy-wide cap-and-trade system to reduce greenhouse gas emissions. However, after discussion, a cloture motion on this bill failed, and the bill was tabled.

While some provisions in energy laws enacted over the past 16 years have led to lower greenhouse gas emissions or addressed climate change directly, other provisions in those same laws have almost certainly resulted in higher emissions. To date, no energy law has had reducing greenhouse gas emissions as the main organizing principle. Energy-related activities are responsible for about 86% of the country's greenhouse gas emissions, and 98% of its carbon dioxide emissions. Climate change policy directed at reducing greenhouse gas emissions must address energy supply and consumption and, thus, be integrated with energy policy. This will be a pivotal challenge to the 111th Congress's and the incoming Administration's anticipated efforts to enact legislation to limit greenhouse gas emissions.

Chapter 4 - This chapter is edited and excerpted testimony by Carol M. Browner before the Committee on Environment and Public Works on April 24, 2007.

Chapter 5 - This chapter is edited and excerpted testimony by David Doniger before the Committee on Environment and Public Works on April 24, 2007.

Chapter 6 - This chapter is edited and excerpted testimony by Peter Glaser before the Committee on Environment and Public Works on April 24, 2007.

Chapter 7 - This chapter is edited and excerpted testimony by Ann R. Klee before the Committee on Environment and Public Works on April 24, 2007.

Chapter 8 - This chapter is edited and excerpted testimony by Honorable William K. Reilly before the Committee on Environment and Public Works on April 24, 2007.

Chapter 9 - This is an edited, excerpted and augmented edition of a United States Congressional Research Service publication, Report Order Code RS22665, dated May 18, 2007.

In: Climate Change Litigation and Law ISBN: 978-1-60876-089-3
Editors: Jean-François Masson

Chapter 1

CLIMATE CHANGE LITIGATION: A GROWING PHENOMENON

Robert Meltz
American Law Division

SUMMARY

The scientific, economic, and political questions surrounding climate change have long been with us. This report focuses instead on a relative newcomer: the legal debate. Though the first court decision related to climate change appeared 18 years ago, such litigation has proliferated in just the past five.

The court cases, decided and pending, arise in seven contexts. The first is the Clean Air Act (CAA). In April, 2007, the Supreme Court held in *Massachusetts v. EPA* that at least as to mobile sources of emissions (cars, trucks), EPA has authority under the CAA to regulate greenhouse gas (GHG) emissions. This decision puts pressure on EPA to move forward as well with regulation of GHGs from stationary sources (powerplants, factories).

Second, litigation under the Marine Mammal Protection Act, and current and likely litigation under the Endangered Species Act, raise the possibility that the impacts of climate change on wildlife may constrain government and private activities associated with GHG emissions.

Third, what about federal energy statutes? A much-publicized decision held that under the Energy Policy and Conservation Act, the United States must monetize the benefits of reduced carbon emissions as part of setting light-truck fuel economy standards. The Outer Continental Shelf Lands Act also has been invoked.

Fourth, various statutes requiring federal government analysis and information dissemination — the National Environmental Policy Act (NEPA), Global Change Research Act, and Freedom of Information Act — have generated climate-change litigation. NEPA suits make up the most numerous category of climate-change litigation. Courts appear to agree that if a plaintiff can establish standing, NEPA can be used to compel agency consideration of the climate change effects of its actions.

Fifth, can common law tort theories such as nuisance be used to force cutbacks in GHG emissions, or payment of damages? All four decisions thus far have denied relief to plaintiffs on lack of standing or political question grounds.

Sixth are the suits challenging state regulation of GHG emissions from motor vehicles as preempted by the federal corporate average fuel economy standards or federal authority over foreign policy. The two rulings thus far have rejected these challenges. Moreover, these issues will be moot unless EPA's denial of California's request for a waiver of federal preemption under the CAA is overturned in court.

Seventh, chiefly with respect to coal-fired power plants, what role will be played by state utilities laws?

Finally, the report discusses international law aspects of a nation's contributions to climate change, and the few international claims made against the United States to date.

INTRODUCTION

The scientific, economic, and political questions surrounding climate change have long been with us. This report focuses instead on a relative newcomer: the legal debate. Though the first court decision related to climate change appeared 18 years ago, the quantity of such litigation has mushroomed in recent years: more than two dozen cases pursuing multiple legal theories are now pending.

The principal court cases, decided and pending, arise in seven contexts — a number that continues to rise. The first and most important context is the Clean Air Act (CAA). In April, 2007, the Supreme Court held in *Massachusetts v. EPA* that at least as to mobile sources of emissions (cars, trucks), EPA has authority under the CAA to regulate greenhouse gas (GHG) emissions.[1]

Second, must global warming effects be considered under the federal wildlife protection statutes, such as the Marine Mammal Protection Act and Endangered Species Act? Third, must they be considered under federal energy statutes, such as the Energy Policy and Conservation Act and Outer Continental Shelf Lands Act? Fourth, may any of several information statutes, particularly the National Environmental Policy Act, be used to compel government analysis of and dissemination of information about climate change? Fifth, can common law tort theories such as nuisance be used by state and private plaintiffs to force cutbacks in GHG emissions, or payment of damages? Sixth, granted that state regulation of GHG emissions from motor vehicles is preempted by the CAA (absent an EPA waiver), is such regulation also preempted by other federal statutory or constitutional law? And seventh, chiefly with respect to coal-fired power plants, what role will be played by state utilities laws?

Sections I through VII of this report address these seven areas of litigation in turn.[2] Most known cases, decided and pending, are mentioned — omitted cases are those that raise climate change issues in only the most marginal way or only implicitly,[3] and some state cases. Looking beyond the domestic lawsuits, Section VIII surveys international-law arguments that might be used to induce GHG emission reductions from the United States and other countries that are major GHG emitters, and the few international-law claims filed against the United States to date. Finally, Section IX offers overall comments.

I. Clean Air Act

Stationary Sources of GHG Emissions

The First EPA General Counsel Memorandum.

Aware that prospects for Senate approval of the Kyoto Protocol were dubious,[4] some Members of Congress became concerned in the late 1990s that the Clinton Administration EPA might seek to regulate GHG emissions in the absence of approval, under either of two claimed authorities. One authority would derive from an argument that even prior to ratification, the Protocol provided some sort of legal basis for emissions restrictions, perhaps citing past treaties signed by the United States that were provisionally implemented prior to going into effect.[5] This possibility provoked a series of enactments barring EPA's use of appropriated funds to implement the Kyoto Protocol in the absence of approval and ratification.[6]

The rest of this section deals with the second issue: EPA's possible authority to regulate GHG emissions independently of the Protocol, under the CAA. This authority has now been confirmed by the Supreme Court, at least as to mobile sources; nonetheless, this report retains from earlier versions some of the historical evolution of the issue for the sake of a fuller appreciation of its complexity.

During hearings on EPA's FY1 999 appropriations, Representative Tom DeLay asked then-EPA Administrator Carol Browner whether the EPA believed it had authority under the CAA to regulate GHG emissions. This led, weeks later, to an EPA General Counsel memorandum,[7] which concluded that CO_2 satisfies the CAA definition of "air pollutant," but that this is only the first step. Before EPA can regulate CO_2 emissions, it must further conclude that CO_2 meets criteria in other CAA provisions requiring the agency to determine that the substance poses harm to public health, welfare, or the environment. This next step EPA declined to take.

At a House hearing in 1999,[8] a panel of legal experts argued the question of EPA's authority to regulate CO_2 under the CAA. A new EPA General Counsel endorsed his predecessor's analysis in the 1998 memorandum, but just as his predecessor, stressed that the EPA's legal analysis was "largely theoretical" since "EPA currently has no plans to regulate carbon dioxide... ."[9] This hands-off position was prompted in part by strong congressional opposition based on uncertainties as to the economic impact of regulating a pollutant as widespread as CO_2. In addition, some in Congress argued that CAA implementation of a CO_2 standard was barred by the aforementioned enactments (appropriation riders) prohibiting implementation of the Kyoto Protocol.[10]

The EPA General Counsel opinion that "air pollutant" includes GHGs held sway until 2003, when that office reversed itself in the context of a petition asking the agency to regulate GHG emissions from mobile sources. This story picks up below ("Mobile Sources of GHG Emissions").

Suits Enforcing the CAA against EPA

Two suits have been filed seeking to compel EPA to regulate GHG emissions from stationary sources.

In *Massachusetts v. Whitman*, filed in 2003, three Northeast states (MA, CT, ME) sought to force EPA to list CO_2 as a "criteria pollutant" under the CAA.[11] They argued that on various occasions, EPA had indicated its belief that CO_2 emissions contribute to climate change. These EPA statements constituted, in the words of CAA section 108,[12] a "judgment [that GHG emissions] cause or contribute to air pollution which may reasonably be anticipated to endanger public health or welfare" and, also per section 108, that such emissions "result[] from numerous or diverse mobile or stationary sources." These prerequisites being satisfied, the suit argued, section 108 required EPA to add CO_2 to its list of "criteria pollutants," then proceed under section 109[13] to develop national ambient air quality standards for CO_2. On September 3, 2003, a few days after EPA's denial of a petition asking the agency to regulate GHG emissions from motor vehicles, the plaintiff states voluntarily dismissed this suit without prejudice, reportedly so as to transfer their energies to a suit challenging the petition denial (leading to the Supreme Court's *Massachusetts v. EPA* decision).

The second suit is *New York v. EPA*, which seeks to compel EPA to issue a New Source Performance Standard (NSPS) for CO_2. Unlike the first case, however, this one will be litigated in the shadow of *Massachusetts v. EPA*, holding that EPA has authority under the CAA to regulate GHGs from *mobile* sources. The burning question is how this litigation seeking to compel EPA regulation of *stationary-source* GHG emissions will be affected.

New York began with an EPA proposal to revise its NSPSs for electric utility and other steam-generating units. Some commenters on the proposed rule argued that EPA must, in addition to the revisions proposed, set NSPSs for GHGs emitted from steam generating units. The commenters pointed to CAA section 111's command that EPA promulgate NSPSs to address emissions from new stationary sources that "cause[], or contribute[] significantly to, air pollution which may reasonably be anticipated to endanger public health or welfare." In promulgating its final rule in February, 2006,[14] however, EPA rejected this demand, saying it lacked authority to set NSPSs for GHGs.

Review of the final rule was sought in the D.C. Circuit.[15] In September, 2006, the court severed the portion of the case dealing with regulation of GHGs, titling it *New York v. EPA*.[16] In September, 2007, a few months after the Supreme Court decision in *Massachusetts v. EPA*, this severed case was remanded to EPA for further proceedings in light of that decision (this section *infra*).

Suits Enforcing the CAA against Stationary Sources

In *Northwest Environmental Defense Center v. Owens Corning Corp.*, environmental groups invoke the CAA citizen suit provision to enforce the act's "new source review" requirement as to GHG emissions.[17] They contend that Owens Corning is constructing a manufacturing plant in Oregon with the potential to emit more than 250 tons per year of harmful gases, without having obtained the required permit.[18] The principal such gas is HCFC- 142b, which plaintiffs contend is a potent GHG. In a preliminary ruling, the court held that plaintiffs have standing, notwithstanding that the climate change impacts of the plant's GHG emissions would be "indirect." Anticipating the Supreme Court's rationale for granting standing in *Massachusetts v. EPA*, the court found that standing was not precluded by the fact that the injury to plaintiffs would be shared with many others, nor because the relief sought would not lead to a *complete* elimination of climate change impacts.

In another proceeding, this time before EPA's Environmental Appeals Board, Sierra Club objected to the issuance of a CAA Prevention of Significant Deterioration (PSD) permit by a state agency for construction of a coal-fired electric power plant. *In re Christian County, LLC*, PSD Appeal No. 07-0 1 (decided January 28, 2008). Sierra Club's argument was that the permit must contain a "best available control technology" (BACT) limit on the plant's CO_2 emissions. The petition was denied because Sierra Club had not raised its argument during the public comment period on the draft permit, but the issue is still alive. Sierra Club is also the petitioner in another case before the Board, *In re Deseret Power Electric Cooperative*, PSD Appeal No. 07-03 (filed October 1, 2007), in which it raised the CO_2 BACT issue during the public comment period and, therefore, preserved the issue for appeal.

In re Deseret involves an EPA-issued PSD permit to construct a new waste- coal-fired utility generating unit at an existing power plant. The PSD BACT issue stems from the fact that the CAA requires a new "major emitting facility" in a PSD area to install BACT "for each pollutant subject to regulation under this act."[19] Sierra Club argues that in light of the *Massachusetts v. EPA* holding that GHGs are "air pollutants" under the act, CO_2 is "subject to regulation" by the CAA. To the contrary, argues EPA, the phrase "subject to regulation" refers only to pollutants that the agency has *specifically* regulated, and since it has yet to issue a regulation limiting CO_2 emissions, BACT for CO_2 is not required here.

Mobile Sources of GHG Emissions

The Section 202 Petition Denial and the Second EPA General Counsel Memorandum

In 1999, 19 organizations petitioned EPA to regulate emissions of GHGs (CO_2, methane, nitrous oxide, and hydrofluorocarbons) from new motor vehicles. The rulemaking petition cited the agency's alleged mandatory duty to do so under CAA section 202(a)(1).[20] That section directs the EPA Administrator to prescribe emission standards for "any air pollutant" from new motor vehicles "which, in his judgment cause[s], or contribute[s] to air pollution which may reasonably be anticipated to endanger public health or welfare."

In 2003, EPA denied the section 202 petition.[21] Much of the agency's rationale followed a new General Counsel memorandum, issued the same day.[22] Contrary to its Clinton Administration precursor, this new OGC memorandum concluded that the CAA does *not* grant EPA authority to regulate CO_2 and other GHG emissions for their climate change impacts.

Massachusetts v. EPA: The Challenge to EPA's Petition Denial

D.C. Circuit Decision

EPA's denial of the section 202 petition prompted a suit in the D.C. Circuit by twelve states (CA, CT, IL, MA, ME, NJ, NM, NY, OR, RI, VT, WA) and others. Opposing the challenge, besides EPA, were ten state intervenors (AK, ID, KS, MI, ND, NE, OH, SD, TX, UT), plus several automobile- and truck-related trade groups.

In 2005, a split panel in *Massachusetts v. EPA* rejected the suit.[23] The two judges supporting rejection, however, did so for different reasons. Judge Randolph, *assuming* that EPA had CAA authority to regulate GHG emissions, spoke to whether EPA properly *exercised its discretion* in choosing not to wield that authority. Recall that CAA section 202(a)(1) directs the EPA Administrator to prescribe standards for any motor vehicle emissions that *"in his judgment"*[24] cause harmful air pollution. Judge Randolph read "in his judgment" broadly to allow EPA consideration of not only "scientific uncertainty" about the effects of GHGs, but also "policy considerations" that justified not regulating. Thus, EPA in his view was entitled to rely, as it did, on such factors as the fact that new motor vehicles are but one of many sources of GHG emissions, resulting in an inefficient piecemeal approach to climate change, and efforts to promote fuel cell and hybrid vehicles. He concluded that EPA had properly exercised its 202(a)(1) discretion in denying the petition for rulemaking. By contrast, Judge Sentelle held that petitioners lacked standing.[25]

In dissent, Judge Tatel asserted that at least one petitioner had standing. Massachusetts, he said, had shown the possibility of injury from global-warming-induced rising sea levels. On the merits, he found first that EPA has the authority under section 202(a)(1) to regulate GHG emissions. Second, he concluded that EPA's 202(a)(1) discretion does not extend to policy considerations, as Judge Randolph held, but relates exclusively to whether the emissions cause harm.

Supreme court decision

In *Massachusetts v. EPA*, the Court ruled 5-4 for petitioners on all three issues in the case — standing, authority, and discretion.[26] First, Massachusetts has standing to bring the claim. Second, EPA has authority to regulate motor vehicle GHGs under section 202, since "air pollutant" includes GHG emissions. And third, the phrase "in his judgment" in section 202 does not allow EPA to exercise discretion against regulating based on policy considerations. The ruling in favor of petitioners was forecast early in the majority opinion by its opening sentences: "A well-documented rise in global temperatures has coincided with a significant increase in the concentration of carbon dioxide in the atmosphere. Respected scientists believe the two trends are related."[27] (Nor did the dissenters dispute this.)

Most of the decision is devoted to standing. The Court found that petitioners had two factors in their favor. First, the CAA specifically authorizes challenges to agency action unlawfully withheld, such as here.[28] A litigant to whom Congress has accorded such a procedural right, said the Court, "can assert that right without meeting the normal standards for redressability and immediacy"[29] — two prerequisites of the standing test. Second, the Court found it "of considerable relevance"[30] that the petitioner injury on which it focused — Massachusetts's loss of shore land from global-warming-induced sea level rise — was that of a sovereign state. States are "not normal litigants for the purposes of invoking federal jurisdiction,"[31] said the Court, noting their quasi-sovereign duty to preserve their territory.

Having described petitioners' favored position in establishing standing, it was surprising that the Court then undertook a traditional standing analysis. As to the first prong of the standing test — whether plaintiff has demonstrated actual or imminent "injury in fact" of a concrete and particularized nature —

the Court homed in on Massachusetts's status as *owner* of much of the shore land being lost to sea level rise. That this injury may be widely shared with other coastal states does not disqualify this injury, said the Court; it is nonetheless concrete.

The second prong of the standing test is causation, requiring that the injury of which the plaintiff complains is fairly traceable to the defendant. EPA did not dispute the existence of a causal relationship between GHG emissions and climate change. It did argue, however, that any reduction in GHG emissions achieved through the current litigation would be too small a portion of worldwide GHG emissions to make a cognizable difference in climate change. In a ruling that may be of benefit to environmental plaintiffs in many contexts, the Court held that even an agency's refusal to take a "small incremental step"[32] that would result in only a modest reduction in worldwide GHG emissions, is enough for standing purposes.

The third and final prong of the standing test is redressability, demanding that the remedy sought by the plaintiff is one that is likely to redress his/her injury. Here, the remedy sought was EPA regulation of GHG emissions from new motor vehicles. The Court found that this remedy satisfied redressability because while it would not by itself reverse climate change, it would nonetheless slow or reduce it. Nor, given the "enormity"[33] of the potential effects of climate change, was it relevant to the Court that the full effectiveness of the remedy would be delayed until existing cars and trucks on the road were largely replaced by new ones.

In contrast with the majority opinion's lengthy discourse on standing, its disposition of the CAA issues in the case is quite brief. On the authority question, the CAA's broad definition of "air pollutant" simply could not, the Court said, be squared with EPA's view that GHGs are not included. The Court rejected EPA's argument that federal laws enacted following enactment of this definition — laws emphasizing interagency collaboration and research — suggest that Congress meant to curtail EPA's power to use mandatory regulations in addressing air pollutants. Nor was the Court impressed with EPA's argument that "air pollutant" in the CAA could not include vehicle GHG emissions because EPA standards for such emissions could be satisfied only by improving fuel economy, a goal assigned to the Department of Transportation under a different statute (the Energy Policy and Conservation Act[34]).

Finally, on the discretion issue, the majority concluded that "in his judgment" refers only to whether an air pollutant "may reasonably be anticipated to endanger public health or welfare." Thus, said the Court, EPA can avoid taking further action in response to the section 202 petition "only if it determines that greenhouse gases do not contribute to climate change or if it provides some reasonable explanation as to why it cannot or will not exercise its discretion." It rejected EPA's stated policy reasons for refusing to regulate GHG emissions, such as its claim that voluntary executive branch programs already provide an effective response to climate change and that unilateral EPA regulation of vehicle GHG emissions could weaken U.S. efforts to persuade developing countries to reduce the GHG intensity of their economies. Such reasons "have nothing to do with whether greenhouse gas emissions contribute to climate change."[35] In short, said the Court, the only question is whether sufficient information exists to make an endangerment finding under section 202.

Accordingly, the Supreme Court reversed the D.C. Circuit opinion and remanded the case to that court for further proceedings.[36] On September 14, 2007, the D.C. Circuit vacated EPA's denial of the section 202 petition and remanded the matter to the agency.

A four-justice dissent by Chief Justice Roberts in *Massachusetts v. EPA* disputed the majority's holding of standing. A dissent by Justice Scalia for the same four justices argued that agency policy preferences may appropriately be considered as part of EPA's decision *whether* to issue a "judgment," conceding that the judgment, *if made*, must be limited to whether vehicle GHG emissions cause endangerment. Justice Scalia also disputed the majority's holding that "air pollutant" in section 202 includes GHGs.

Implications of supreme court decision

The Court's decision leaves EPA with three options: make a finding that motor vehicle GHG emissions may "endanger public health or welfare" and issue emissions standards; make a finding that such emissions do not satisfy that prerequisite; or decide that climate change science is so uncertain as to preclude making a finding either way (or cite some other "reasonable explanation" why it will not exercise its discretion either way).[37] As to the state of climate change science, the Court's focus on the policy reasons EPA gave as part of exercising its "judgment" obscures the fact that the agency's rejection of the petition stemmed in part from expressions of scientific uncertainty in a 2001 National Research Council report on the science of climate change. Whether scientific reports since the petition rejection in 2003 have foreclosed the scientific-uncertainty rationale is beyond the scope of this report.

The EPA Administrator did say after the decision that although it bars EPA use of policy considerations as a basis for *denying* the petition, it left open whether the agency can invoke them when actually *writing* the regulations, should the agency make an endangerment finding.[38] In this regard, it should be noted that CAA section 202 does not explicitly impose any stringency or other criteria on GHG emission standards promulgated under the section. Given the wide latitude EPA has in setting section 202 standards for GHGs, the possibility presents itself that EPA, following an endangerment finding, could set voluntary standards, or standards pegged to the CAFE standards for fuel economy, or standards that must be complied with only after the President certifies that developing nations have put adequate GHG emission limits into effect. We offer no opinion here as to the legal viability of these options.

On May 14, 2007, the President asked EPA to have CAA regulations limiting vehicle GHG emissions in place by the end of 2008 and to use the President's 2007 State of the Union proposal for raising the CAFE standards as a guide.[39] As late as early December, 2007, EPA was consistently stating that it intended to issue proposed regulations by the end of 2007. However, the enactment of the Energy Independence and Security Act in December, 2007,[40] with its increase in CAFE standards, led EPA to back off from any firm deadline for issuance of mobile-source GHG emission standards. In January, 2008, letters from the *Massachusetts v. EPA* petitioners asked the EPA Administrator when he intended to act,[41] prompting EPA to respond that it did not have a specific time line. Most recently, EPA announced by letter that it will issue during spring, 2008, an advance notice of proposed rulemaking (ANPR) addressing the full range of *Massachusetts v. EPA*'s ramifications throughout the CAA, including the section 202 standards[42] — on the ground that regulation under section 202 could automatically result in other CAA regulations applying to stationary sources and other mobile sources. In response, the *Massachusetts v. EPA* petitioners requested the D.C. Circuit to order EPA to comply with the Supreme Court's remand and the Circuit's mandate within 60 days (choosing one of the three options noted earlier), arguing that the proposed ANPR extended well beyond the narrow section 202 endangerment issue in the case.[43]

As the EPA letter confirms, the Court's ruling in *Massachusetts v. EPA* has many implications beyond its four corners.

On the substantive (non-standing) side, the Court's ruling upholding CAA coverage of GHG emissions from mobile sources likely will improve the prospects of litigation seeking to have EPA restrict GHG emissions from stationary sources as well. The stationary-source provisions of the CAA use terms similar to that of section 202 — in particular, "air pollutant," "in his judgment," and "endanger."[44] As the above discussion of *New York v. EPA* indicates, such an effort to compel EPA regulation of stationary source GHGs is already underway as to NSPSs. Further, if EPA sets a national ambient air quality standard for CO_2, GHGs would be covered under the CAA's new source review permitting program for major emitting facilities and modifications in Prevention of Significant Deterioration areas.[45] Presumably, best available control technology for CO_2 emissions would then have to be installed on such facilities.[46]

On the mobile-source side, *Massachusetts v. EPA* is expressly relied upon by California in three rule-making petitions asking EPA to regulate GHG emissions from mobile sources other than those at issue in that decision — in particular (1) ocean-going vessels under CAA section 213 (a)(4), which authorizes federal emission standards for "new nonroad engines and new nonroad vehicles" (petition submitted October, 2007);[47] (2) aircraft under CAA section 231 (submitted December, 2007);[48] and (3) new nonroad vehicles and engines excluding aircraft, locomotives, and vessels (i.e., farm and construction equipment, lawn and garden equipment, RVs, etc.) under CAA sections 202 and 213(a)(4) (submitted January, 2008).

Non-substantively, the Supreme Court's ruling on standing will likely be pivotal to the fortunes of plaintiffs in other climate change litigation. The question will be the extent to which the Court's finding of standing was contingent, as it obliquely suggested, on the existence of a state-sovereign plaintiff[49] and the presence in the CAA of an explicit provision allowing the filing of administrative petitions.

Ironically, the "environmental win" in *Massachusetts v. EPA* has thwarted the environmental side in a climate-change-related nuisance case, and may do so in others. One court used the decision as peripheral support for dismissing a nuisance action on "political question" grounds, reasoning that the Supreme Court has now found authority over GHG emissions to reside in the Federal Government.[50] In the future, the decision may also undermine federal common law claims, on the argument that Congress intended to leave no room for courts to develop overlapping federal common law restricting GHG emissions.

II. Wildlife Protection Statutes

Marine Mammal Protection Act

The Marine Mammal Protection Act (MMPA)[51] bars the taking of marine mammals, with exceptions. One exception is for "incidental takings" by specified activities.[52] It provides that persons "engage[d] in a specified activity (other than commercial fishing) within a specified geographical region" may request the Secretary of the Interior or Commerce to authorize, for up to five years, the incidental, but not intentional, taking of small numbers of marine mammals. The Secretary must grant the authorization if he/she makes certain findings — including that the effect of the incidental take will be "negligible" — and promulgates rules setting out permissible methods of taking by the specified activity.

In *Center for Biological Diversity v. Kempthorne*, No. 07-CV-00141 (D. Alaska), *transferred from* No. 07-CV-00894 (N.D. Cal. filed February 13, 2007), environmental groups challenge one such "incidental taking" rule promulgated by the Fish and Wildlife Service (FWS), Department of the Interior. The rule authorizes the incidental take of polar bears and Pacific walrus for five years resulting from oil and gas activities in the Beaufort Sea and adjacent coastal areas of the Alaska north slope.[53] Plaintiffs argue that the rule violates the MMPA and thus is arbitrary and capricious by permitting more than a "negligible" impact on the species, based on the combined impact of oil-and-gas activities and climate change.[54] (This lawsuit also contains a National Environmental Policy Act claim, discussed in Section IV.)

Endangered Species Act

Under the Endangered Species Act (ESA),[55] animals (and plants) may be listed as endangered or threatened. As to a listed species, two of the act's provisions are particularly relevant. Section 9 makes it unlawful to "take" a member of a listed endangered species,[56] and has been extended by regulation to most threatened species.[57] Exceptions from the take prohibition are possible, as through incidental take permits. The other provision, section 7, demands that each federal agency "insure that any action authorized, funded, or carried out by such agency ... is not likely to jeopardize the continued existence of any endangered species or threatened species or result in the destruction or adverse modification of [designated critical habitat] of such species....”[58] To achieve this goal, section 7 directs a federal agency, upon finding in a biological assessment that a proposed agency action is “likely to affect” a listed species, to consult with the appropriate wildlife agency — the FWS or National Marine Fisheries Service (NMFS). This is called “section 7 consultation.” Then, the FWS or NMFS prepares a “biological opinion” concluding either that the proposed action would not violate the no-jeopardy/adversemodification mandate, or that it would violate the mandate, in which case FWS or NMFS must suggest “reasonable and prudent alternatives” that would not violate the mandate.

In *Natural Resources Defense Council v. Kempthorne*, 506 F. Supp. 2d 322 (E.D. Cal. 2007), environmental and sport fishing groups attacked the FWS biological opinion prepared for the 2004 Long-Term Central Valley Project and State Water Project Operations Criteria and Plan and certain related future actions. The biological opinion concluded that project operations would not jeopardize the continued existence of the Delta smelt, a threatened species, or adversely modify its designated critical habitat — that is, would not violate ESA section 7. The court, however, held that the biological opinion was arbitrary and capricious in ignoring data about climate change that may adversely affect the Delta smelt and its habitat. The court observed, for example, that the opinion was based on the assumption that the hydrology of the waters affected by the 2004 plan would follow historical patterns for the next 20 years, an assumption that studies on the potential effects of climate change on water supply reliability did not support.

A companion case pending before the same judge, *Pacific Coast Federation of Fishermen's Associations/Institute for Fisheries Resources v. Gutierrez*, No. 1:06- CV-00245 (E.D.Cal. filed August 9, 2005), challenges the NMFS biological opinion prepared in connection with the same project for various salmon and trout species. (A NEPA claim in this suit was dismissed. 2007 Westlaw 1752289 (E.D. Cal. June 15, 2007).)

It may be only a matter of time before several more ESA cases are filed. In an effort spearheaded by the Center for Biological Diversity, multiple petitions have been filed to have animals listed as endangered or threatened due in various degrees to climate change impacts on their habitat. Should such species then be listed and their critical habitats designated, the Center is likely to test in court whether substantial GHG sources run afoul of protections afforded those species by the ESA.

Only one climate-change-related proposal to list a species has reached the actual listing stage thus far — that by NMFS bestowing threatened status on the staghorn coral and elkhorn coral.[59] Considerably more attention to ESA implications, however, has been directed at the January, 2007 FWS proposal to list the polar bear as threatened, in fulfillment of a settlement agreement with environmental groups including the Center for Biological Diversity.[60] The polar bear proposal ties in with widely publicized studies as to the impact of climate change on the Arctic — particularly, the loss of sea ice required by polar bears as habitat.[61] In addition to the coral and polar bear, the Center has petitioned the FWS to list as either endangered or threatened Kittlitz's murrelet, a seabird (2001), 12 species of penguins (2006),[62] the American pika, an alpine mammal (2007), the ashy storm-petrel, another seabird (2007), the ribbon seal (2007), and the Pacific walrus (2008). In each instance, the Center asserts global warming to be a cause, principal or otherwise, of the species' plight.

If the polar bear, for example, is listed, would starting up a new fossil-fuel-fired power plant violate section 9 — cause a "take" — through the effects of its GHG emissions, via climate change, on polar bear habitat?[63] Notable here is that "take" is statutorily defined to include "harm" to a member of a listed species, and "harm," in turn, is defined by regulation to include certain "significant habitat modification[s] or degradation[s]."[64] The crux, presumably, is whether the causal link between the power plant's GHG emissions and the effect on the species habitat is sufficiently direct and substantial to constitute a "take," a question beyond the scope of this report. If a take is found, the power plant would require an incidental take permit to operate, such permit likely containing restrictions on the amount of GHGs that could be emitted. Likewise, would a federal agency issuing a permit for powerplant construction have to initiate section 7 consultation?

To be sure, the ESA and its regulations do contain mechanisms — e.g., incidental take permits, "reasonable and prudent alternatives," and "special rules" for threatened species — that can ease the act's constraints on human activities affecting listed species.

III. ENERGY STATUTES

Energy Policy and Conservation Act

In *Center for Biological Diversity v. National Highway Traffic Safety Administration*, 508 F.3d 508 (9th Cir. 2007), 11 states (CA, CT, ME, MA, NJ,

NM, NY, OR, RI, VT, MN), four environmental groups, and others attacked a 2006 rule promulgated by the National Highway Traffic Safety Administration (NHTSA) under the Energy Policy and Conservation Act (EPCA). The rule established corporate average fuel economy (CAFE) standards for light-duty trucks — defined by NHTSA to include many SUVs, vans, and pickup trucks — in model years 2008 through 2011.

EPCA says that the light-truck CAFE standard shall be the "maximum feasible" standard that manufacturers can achieve in a given model year.[65] The court found that even assuming NHTSA may use a cost-benefit analysis to determine the "maximum feasible" standard, it was arbitrary and capricious not to include in the analysis the benefit of carbon emissions reduction — calling this "the most significant benefit of more stringent CAFE standards."[66] NHTSA had argued, for example, that the wide range of values put forward in studies as to how the benefits of reduced GHG emissions should be monetized justified placing no value on that benefit in its cost-benefit analysis. The court countered that while there is indeed a range of values in the studies, they are all greater than zero. Accordingly, the court remanded the CAFE standard to NHTSA for the agency to include a monetized value for carbon emission reduction in its analysis of the proper CAFE standard. (There was also a climate-change-related NEPA claim in this lawsuit, discussed in Section IV.)

Outer Continental Shelf Lands Act

In a petition for review filed with the D.C. Circuit, the Center for Biological Diversity challenges the June, 2007 approval by the Secretary of the Interior of the Outer Continental Shelf Oil and Gas Leasing Program 2007-2012. *Center for Biological Diversity v. U.S. Dep 't of Interior* [sic], No. 07-1247 (D.C. Cir. filed July 2, 2007). Plaintiff alleges that the Secretary violated the Outer Continental Shelf Lands Act[67] by failing to disclose or analyze the environmental and economic impacts from "the greenhouse gas emissions that would result from use of oil and gas produced as a result of the [Program]."[68] This language indicates that it is not the GHG emissions from the oil and gas production itself that is at issue, but rather the GHG emissions resulting from the "use" of that oil and gas in cars, powerplants, or wherever. (There was also a climate-change-related NEPA claim in this lawsuit, discussed in Section IV.)

IV. INFORMATION STATUTES

Much of the climate change litigation is based on statutory requirements that the government generate, compile, or disclose information.

National Environmental Policy Act

To be sure, the National Environmental Policy Act (NEPA) is more than just an information statute, declaring as it does a sweeping policy that the federal government must consider the environmental impacts of its actions. However, NEPA ensures that such environmental consideration will occur chiefly through the production of information, in the form of environmental assessments and environmental impact statements, and does not require that an agency choose from among its options the one with the least environmental impact.

The NEPA cases involving climate change represent the oldest and most numerous category of climate change litigation. Again, not all cases are mentioned in this report.[69]

The dominant issue has been whether plaintiffs have standing to sue — as mentioned, an issue on which plaintiffs may be helped by the 2007 Supreme Court decision in *Massachusetts v. EPA*. Thus, all the standing issues discussed here should be viewed through the prism of that decision. The standing determination has been particularly difficult in the context of NEPA, which confers only a *procedural* right (having a federal agency prepare an adequate environmental impact statement (EIS)), not a *substantive* right (having the agency select a particular course of action after preparing the EIS). Where courts have found standing and reached the merits, they have usually accepted that climate change impacts in the proper circumstances are a required consideration in an EIS.[70]

District of Columbia Circuit

Standing barriers have proved particularly daunting in the D.C. Circuit, thus it is here that *Massachusetts v. EPA* may have its greatest effect. In the first significant climate change case, *City of Los Angeles v. National Highway Traffic Safety Admin.*, 912 F.3d 478 (D.C. Cir. 1990), the city attacked a NHTSA decision not to prepare an EIS when it set the corporate average fuel economy standard at 26.5 mpg for model year 1989 passenger cars — below the statutory default setting of 27.5 mpg.[71] A majority of the D.C. Circuit panel concluded that petitioners had standing based on their argument that had NHTSA done an EIS considering the climate change impacts of its one mpg rollback, the agency might have rejected it. This provided the requisite causal nexus, said the majority, between NHTSA's decision not to do an EIS and climate change. In dissent, however, one judge argued that Article III demanded a more precise causal showing, with clear proof of a nexus between the agency action and harm to the petitioners. On the merits, one judge in the majority concluded that NHTSA had "inadequately explained why the admitted increase in carbon dioxide is insignificant within the context of the environmental harm posed by global warming."[72] She would have remanded NHTSA's NEPA decision but left the rollback in place in the meantime. Because the other majority judge ruled for the agency, however, the petition was denied.

The plaintiff-friendly *City of Los Angeles* standard for finding global-warming-based standing was to prove short-lived. Six years later, a divided D.C. Circuit declared *en banc* that to obtain standing, a procedural-rights plaintiff must show not only that the government omitted a required procedure, but that it is substantially probable that the procedural omission *will cause a particularized injury to the plaintiff*[73] — adopting the dissenter's position in that case. To the extent *City of Los Angeles* dispensed with the second, causation-of-a-particularized-plaintiff-injury requirement, it was expressly overruled. Still later court decisions, however, have cast doubt on this strict standard.[74]

In *Foundation on Economic Trends v. Watkins*, 794 F. Supp. 395 (D.D.C. 1992), the standing bar was raised during, rather than after, the litigation. Plaintiffs claimed that NEPA required the Secretaries of Energy, Agriculture and the Interior to evaluate the effect on climate change of 42 actions and programs under their authority. Plaintiffs' standing argument was based on "informational standing," under which failure to do an EIS discussing possible climate change impacts satisfies the injury requirement of standing merely by harming plaintiffs' programs for disseminating information about climate change to the public. In so arguing, plaintiffs relied on a line of D.C. Circuit decisions going back two decades. Unfortunately for them, however, informational standing was limited by the D.C. Circuit during the pendency of their suit. An amended complaint by the individual plaintiff, arguing that his expected use of his oceanfront cottage may be curtailed if oceans rise from climate change, was also rejected. Among other things, said the court, the plaintiff had not met the causation requirement of standing in that he had not related the environmental harm he predicted to any of the 42 challenged agency actions. "[T]here is no 'global warming' exception to the standing requirements of Article III or the [Administrative Procedure Act],"[75] it asserted.

In a suit described in Section III, *Center for Biological Diversity v. U.S. Dep 't of Interior* [sic], No. 07-1247 (D.C. Cir. filed July 2, 2007), plaintiff charges that the Secretary of the Interior failed to analyze in the EIS for his five-year Outer Continental Shelf leasing program (1) the GHG emissions resulting from the use of the oil and gas produced under the program, and (2) the effects of global warming on the resources affected by the program "including, but not limited to, polar bears, walrus, and corals."

In *Montana Environmental Information Center v. Johanns*, No. 07-CV-01 331 (D.D.C. filed July 23, 2007), challenge is made to the Department of Agriculture's Rural Utility Service's use of low-interest loans to help finance the construction of at least eight new coal-fired powerplants. The charge is that the EIS for one plant is deficient because it fails to consider the cumulative impacts of GHG emissions from the eight new plants.

Ninth Circuit

The standing barriers in the D.C. Circuit seem to be attenuated in the Ninth Circuit where, as far as research reveals, plaintiffs raising climate change claims in NEPA suits have yet to encounter standing problems.

In 2002, environmental groups sued the Overseas Private Investment Corp. (OPIC) and Export-Import Bank of the United States alleging continued failure to comply with NEPA. These federal agencies provide insurance, loans, and loan guarantees for overseas projects, or to U.S. companies that invest in overseas projects. Plaintiffs alleged that these overseas projects include oil and gas extraction and refining, and power plants, which together result in the annual emission of billions of tons of GHGs, causing climate change in the United States.

In 2005, the district court held that plaintiffs had standing, given what it saw to be the relaxed standards in the Ninth Circuit for showing standing in cases alleging procedural violations — here, failure to prepare an EIS.[76] *Friends of the Earth v. Mosbacher*, 2005 Westlaw 2035596 (N.D. Cal. 2005). It is "reasonably probable," said the court, that emissions from projects supported by the defendants will threaten plaintiffs' concrete interests. In 2007, the court reached the merits, holding on summary judgment motions that defendants need not prepare a programmatic EIS for the energy projects they finance, and that neither side had shown, as a matter of law, that energy projects specifically listed in the complaint are or are not "major Federal actions" requiring an EIS. 488 F. Supp. 2d 889 (N.D. Cal. 2007).

In *Border Power Plant Working Group v. Dep 't of Energy*, 260 F. Supp. 2d 997 (S.D. Cal. 2003), plaintiff challenged the environmental assessment accompanying applications for permits and federal rights of way to build electricity transmission lines connecting new power plants in Mexico with the power grid in Southern California. In part because four of its members were seen to have procedural standing, the plaintiff organization was held to have organizational standing.[77] The court's standing discussion made no mention of climate change, however, perhaps because climate change was only a small part of plaintiff's case. On the merits, the court agreed with plaintiff that the environmental assessment was legally inadequate because, among other things, it failed to discuss CO_2 emissions from the powerplants and "[t]he record shows that carbon dioxide ... is a greenhouse gas."[78]

The most recent NEPA/climate change decision by the Ninth Circuit, *Center for Biological Diversity v. NHTSA*, 508 F.3d 508 (9th Cir. 2007), offers a deja vu to *City of Los Angeles*, discussed earlier in this section. Both cases involve a NHTSA rule setting a corporate average fuel economy (CAFE) standard — this time, for light-duty trucks (model years 2008-20 11)[79] — and in both cases, the agency did no EIS. Petitioners include 11 states (CA, CT, ME, MA, NJ, NM, NY, OR, RI, VT, MN) and four environmental groups. In sharp contrast with earlier NEPA/climate-change decisions, the United States in this case did not contest standing and the court decision does not mention it.

On the merits, the court held that NHTSA's environmental assessment for its CAFE rule, finding no significant impact, was inadequate owing to, among other things, its analysis of the rule's cumulative impacts from GHG emissions. Said the court: "The impact of greenhouse gas emissions on climate change is precisely the kind of cumulative impacts analysis that NEPA requires agencies to conduct."[80] Nor did the Energy Policy and Conservation Act, the statute authorizing CAFE standards, limit NHTSA's duty to assess environmental impacts such as climate change. More specifically, while NHTSA's assessment indicated the expected amount of CO_2 emitted by light-duty trucks under the new CAFE standard, it failed to "evaluate the 'incremental impact' that these emissions will have on climate change ... in light of other past, present, and reasonably foreseeable actions such as other light truck and passenger automobile CAFE standards."[81] Finally, the court invoked the well-settled principle that an EIS must be prepared if substantial questions are raised as to whether a proposed project *may* have significant environmental impact, and held that petitioners' evidence raised the necessary level of doubt. Thus, the court ordered preparation of a full EIS. (There was also a climate change-related Energy Policy and Conservation Act claim, discussed in Section III.)

In *Center for Biological Diversity v. Kempthorne*, No. 07-CV-00141 (D. Alaska), *transferred from* No. 07-CV-00894 (N.D. Cal. filed February 13, 2007), environmental groups challenge a Fish and Wildlife Service "incidental taking" rule. As described in Section II, the rule authorizes the incidental take of polar bears and Pacific walrus by oil and gas activities in the Beaufort Sea and adjacent coastal areas of the Alaska north slope.[82] Plaintiffs challenge the environmental assessment and finding of no significant impact, charging that the Service put out the rule "without seriously analyzing the effects of climate change on them or their habitat." The accusation is not that the oil and gas activities themselves contribute to climate change, but that direct harms to polar bears and walruses from those activities will be exacerbated by climate change impacts on the Arctic that are already stressing those species.

Eighth Circuit

In *Mid States Coalition for Progress v. Surface Transportation Bd.*, 345 F.3d 520 (8th Cir. 2003), petitioners disputed the adequacy of an EIS prepared by the Surface Transportation Board to accompany its approval of a railroad's proposal to construct new rail and upgrade existing rail. The proposed rail line was to provide a less expensive route by which low-sulfur coal in Wyoming's Powder River Basin could reach powerplants, and thus might be expected to increase coal consumption and its attendant effects. In this regard, the court noted that CEQ's NEPA regulations require that EISs cover both direct and indirect effects of proposed actions.[83] It concluded by finding it "irresponsible" for the Board to approve such a large project without first examining the possible effects of an increase in coal consumption — apparently, the opinion suggests (but does not explicitly say), including climate change.[84]

In *Ranchers Cattlemen Action Legal Fund v. Conner*, No. 07-CV-0 1023 (D. S.D. filed October 24, 2007), plaintiffs challenge Department of Agriculture regulations easing restrictions on the import of live cattle and edible bovine products from "minimal risk" Mad Cow Disease regions (Canada). Plaintiffs assert that the environmental assessment was inadequate because it did not analyze the increased GHG emissions from the transportation of the cattle into the United States.

State NEPAs

A few GHG-related suits also have been filed under state "little NEPAs" — state laws requiring state (and sometimes local) agencies to consider the environmental impacts of their proposed actions, just as the federal NEPA does for federal agencies.[85] For example, in *General Motors Corp. v. California Air Resources Bd.*, No. 05-02787 (Cal. Sup. Ct. filed September 2, 2005), two car manufacturers claimed that the Board's adoption of California's GHG emission standards involved delayed and inadequate compliance with the state's NEPA-type law. This suit offers as a prime reason for environmental analysis the argument that GHG emissions regulation has, in addition to a possible benefit, some environmental downsides. In particular, it contends that restriction of GHG emissions may cause an increase in new-vehicle sticker prices and a consequent decrease in the rate at which old, higher-emissions vehicles are retired from use.

State-NEPA cases were filed in April, 2007 by conservation groups and the California attorney general to require San Bernardino County, the largest county in the United States by area, to address climate change in its General Plan update.[86] In August, 2007, California settled its lawsuit, the County agreeing to prepare a Greenhouse Gas Emissions Reduction Plan and adopt other measures, but the conservation groups' suit remains. In broaching the vast realm of local land use plans, these cases portend a major new front in climate change litigation, particularly in states that require environmental impact analysis.

Global Change Research Act

The Global Change Research Act of 1990 (GCRA)[87] commands the President to create an interagency United States Global Change Research Program to improve understanding of "global change." Global change is defined broadly by the GCRA to include all changes in the global environment "that may alter the capacity of the Earth to sustain life." Thus, the statute includes, but goes beyond, climate change.[88] The Program is to be implemented by a National Global Change Research Plan, with regular scientific assessments that evaluate the findings of the Program. The GCRA demands that revised Research Plans be submitted to the Congress at least every three years,[89] with the last one having been submitted July, 2003. The statute further demands that scientific assessments be submitted to the

President and Congress not less often than every four years,[90] with the only assessment to date submitted October, 2000.

On these undisputed facts, the district court in *Center for Biological Diversity v. Brennan*, 2007 Westlaw 2408901 (N.D. Cal. August 21, 2007), had little difficulty finding that the Administration had unlawfully withheld action it was required to take. It ordered defendants to publish a summary of the revised proposed Research Plan no later than March, 2008, with submission to Congress no later than 90 days thereafter.[91] The court further ordered the scientific assessment to be produced by May, 2008. It should be noted in passing that the large bulk of this opinion is devoted not to the foregoing violation and remedy, but to threshold matters: standing (finding procedural rights injury and informational injury, associational standing, and Administrative Procedure Act standing) and a motion to intervene by two Members of Congress (denied).

Freedom of Information Act

The Freedom of Information Act (FOIA)[92] mandates that documents in the possession of federal executive branch agencies are to be disclosed to the public upon request, unless covered by a FOIA exemption.

In May, 2006, Citizens for Responsibility and Ethics in Washington (CREW) invoked FOIA to request from the Council on Environmental Quality (CEQ) all of its records relating to the causes of climate change, from January 20, 2001, to October 26, 2006. Though CEQ produced many documents, CREW sued under FOIA seeking a court order that CEQ release all records responsive to its request. *Citizens for Responsibility and Ethics in Washington v. Council on Environmental Quality*, No. 1 :07CV00365 (D.D.C. filed February 20, 2007). In a fall 2007 filing with the court, the parties noted that CEQ had produced 16,000 pages of documents to CREW by May, 2007, and was about halfway through processing the remaining 27,000 pages.[93]

This lawsuit parallels allegations that political appointees at CEQ have edited many of the agency's reports to minimize the danger and human causes of climate change. In July, 2006, the House Committee on Government Reform[94] requested that CEQ provide documents and communications relating to the agency's edits of climate change materials, its efforts to influence the statements of government scientists, its communications with federal agencies and nongovernmental parties regarding climate change, and so on.

V. Common Law Tort

The widely diverse injuries predicted from climate change mean that a comparably diverse spectrum of plaintiffs and defendants could become involved in common law tort litigation based on such injuries. Possible plaintiffs include property owners (farmers dealing with reduced rainfall, owners of oceanfront homes dealing with rising sea level or increased storm activity), nonowner users of natural resources (ski resort operators, commercial fishermen), and state attorneys general bringing private or public nuisance claims (the former for injury to state-owned land, the latter on behalf of the state's citizenry to protect public health and well-being). Possible defendants include the companies that produce the fossil fuels whose combustion produces GHG emissions, entities that emit GHGs (chiefly fossil-fuelfired powerplants, but many other sources also), and companies that manufacture or market products whose use creates GHG emissions (chiefly the automakers).[95]

Several of these potential plaintiff and defendant categories are represented in the five climate-change-related tort cases known to be filed thus far (four discussed in the following text, one in footnote 105). Thus far, all of those tort actions that have produced court decisions have failed, either due to lack of standing or the political question doctrine, or both. Three are on appeal, however.

Nuisance

In 2004, eight states (CA, CT, IA, NJ, NY, RI, VT, WI) and New York City sued five electric utility companies.[96] *Connecticut v. American Electric Power Co.*, Civ. No. 04 CV 05669 (S.D.N.Y. filed July 21, 2004). These defendants were chosen as allegedly the five largest CO_2 emitters in the United States, through their fossilfuel-fired electric powerplants. Invoking the federal and state common law of public nuisance,[97] plaintiffs seek an injunction requiring defendants to abate their contribution to the nuisance of climate change by capping CO_2 emissions and then reducing them. Plaintiffs sue both on their own behalf to protect state-owned property (e.g., the hardwood forests of the Adirondack Park in New York), and as *parens patriae* on behalf of their citizens and residents to protect public health and well-being.

On the same day, three nongovernmental organizations filed a similar suit against the same defendants, in the same court, adding a private nuisance claim.[98] *Open Space Institute v. American Electric Power Co.*, No. 04 CV 05670 (S.D.N.Y. filed July 21, 2004). They seek to protect land owned and preserved by them in the state of New York.[99] This suit was consolidated with the state suit.

In a series of motions, defendants sought to have these actions dismissed on a wide spectrum of threshold grounds. Though the case has now been decided by the trial court on a single threshold issue, it is worth reviewing the grounds advanced in these motions because they may reappear later, in this or other private GHG litigation.[100] To reiterate, many of these grounds typify the difficulties encountered when one seeks to address through private litigation a ubiquitous, long-term environmental problem to which countless parties contribute.

In a dismissal motion, some defendants argued there is no federal common law cause of action for climate change. Creating such federal common law, they argued, runs afoul of Supreme Court directives that federal courts do so only in limited areas — especially where, as with climate change, the problem at issue has sweeping implications. Even assuming a viable federal common-law nuisance theory, they continued, Congress's enactment of a comprehensive scheme of air pollution regulation in the CAA displaces federal court authority in this area. Defendants also challenge plaintiffs' standing to sue. Plaintiffs, they argued, have not demonstrated the "injury in fact" requisite of standing because they allege only injuries from climate change in the indefinite future. Nor, said these defendants, have plaintiffs shown "causation" because they do not allege that defendants' conduct will directly cause the consequences of climate change — especially since defendants' collective emissions are admitted to be less than 2-1/2% of the global total from human activities.[101] As mentioned, the viability of these federal common law of nuisance and no-standing arguments by defendants may be significantly affected — the displacement argument helped, the others hurt — by *Massachusetts v. EPA*.

A second motion sought dismissal for lack of personal jurisdiction. These movants alleged that although their powerplants are located in 20 states, none is in New York (where the case was filed) or any other plaintiff state except Wisconsin.

A final motion to dismiss asserted that to the limited extent a federal common law claim to abate an interstate nuisance has been recognized, it has been limited to actions brought by state entities. Nor, said defendants, can plaintiffs assert public nuisance, because they have not alleged special injury to their properties, or private nuisance, because they have not alleged substantial harm.

As indicated, the dismissal motions in *Connecticut* and *Open Space Institute* have now been ruled on by the district court,[102] which dismissed the cases on political question grounds. This judicial doctrine requires a court to look into "the appropriateness under our system of government of attributing finality to the action of the political departments [i.e., the legislative and executive branches] and also the lack of satisfactory criteria for a judicial determination...."[103] One situation judicially recognized as pointing to a political question, hence dismissal of the action, is "the impossibility of deciding [the case] without an initial policy determination of a kind clearly for nonjudicial discretion."[104] This situation, said the court, perfectly fit the GHG cases, touching as they do on so many areas of national and international policy. As a political question, the court believed the climate change issue in these suits to be for the legislature, not the courts, to resolve. Very possibly, the amorphousness of nuisance law, giving the court little guidance in resolving these cases, may have hurt the plaintiffs' cause. (These cases are now on appeal to the Second Circuit.[105])

A second nuisance action was filed in 2006 by California against several automobile manufacturers based on the alleged contributions of their vehicles to climate change impacts in the state. The suit alleges that these impacts constitute a public nuisance under federal or state common law, and seeks monetary damages (plaintiffs in *Connecticut* seek injunctive relief). The district court dismissed the suit on the same political-question rationale as in *Connecticut* — namely, "the impossibility of deciding without an initial policy determination of a kind clearly for nonjudicial discretion." *California v. General Motors Corp.*, 2007 Westlaw 2726871 (N.D. Cal. September 17, 2007). The need for an "initial policy determination" by the political branches was supported, in the court's view, by the complexity of the climate change issue, the need for political guidance in divining what is an "unreasonable" interference with the public's rights (the definition of a public nuisance), and the existing debate over global warming in the political branches. Ironically, the environmental "win" in *Massachusetts v. EPA* was cited by the court against the state, both because that decision found authority over GHG emissions to lie with the federal government and because it recognized a state's standing to press its grievances at the federal level. (An appeal to the Ninth Circuit is pending.)

Most recently, a native village on the Alaska coast sued certain oil and energy companies, claiming that climate change is destroying the village through the melting of Arctic sea ice that formerly protected it from winter storms. *Native Village of Kivalina v. Exxonmobil Corp.*, No. 08-cv-01138 (N.D. Cal. filed February 26, 2008). Indeed, the complaint asserts, "[t]he U.S. Army Corps of Engineers and U.S. Government Accountability Office have both concluded that the village must be relocated due to global warming...." The village invokes the federal common law of public nuisance, state statutory or common law of private and public nuisance, and a civil conspiracy claim. The conspiracy claim asserts that some of the defendants have engaged in agreements to participate in the intentional creation or maintenance of a public nuisance — that is, global warming — by misleading the public as to the science of global warming.

Negligence, etc.

Owners of Mississippi property damaged by Hurricane Katrina sued certain oil, coal, and chemical companies, alleging a multistep chain of causation: the companies emitted GHGs, which contributed to global warming, which made the waters of the Gulf of Mexico warmer, which caused Hurricane Katrina to become more intense as it passed over the Gulf than it would otherwise have been, which increased the harm to plaintiffs' property caused by the hurricane. Plaintiffs asserted various state-law tort claims, including negligence, nuisance (public and private), and trespass, and seek compensatory damages; they request punitive damages for gross negligence. Further, they claimed fraud and conspiracy to commit fraud, alleging that the oil and coal companies disseminated misinformation about global warming. Finally, plaintiffs made claims against their home insurance companies (e.g., breach of fiduciary duty claim for misrepresenting policy coverage, and violation of a state consumer-protection act) and their mortgage companies (arguing that they may not claim sums owed by plaintiffs for the value of the mortgaged property that was uninsured).

The district court, sitting in diversity, dismissed the action for lack of plaintiff standing. *Comer v. Murphy Oil USA, Inc.*, Civ. Action No. 1 :05-CV-43 6-LG-RHW (S.D. Miss. August 30, 2007). With regard to certain defendants, the court also found plaintiffs' claims nonjusticiable under the political question doctrine — as in the decisions above where nuisance was the sole legal theory advanced. (An appeal to the Fifth Circuit is pending.)

VI. FEDERAL PREEMPTION

Stationary Sources of GHG Emissions

The question of whether federal law preempts state regulation of GHG emissions arises chiefly in connection with *mobile* sources. With limited exceptions, the CAA disclaims any intention to preempt state air pollution controls on *stationary* sources.[106] And the Energy Policy and Conservation Act does not set fuel economy standards for other than mobile sources, so it too would be unlikely to preempt state regulation of stationary sources. However, some have asserted that state regulation of stationary-source GHGs is preempted as contrary to the federal government's authority over foreign policy — an argument being pressed, so far unsuccessfully, in litigation attacking state regulation of *mobile-source* GHG emissions (see below). The most prominent state enactment limiting GHG emissions from stationary sources is that of California, which so far has not been challenged.[107]

Mobile Sources of GHG Emissions: CAA Preemption

The picture is quite different for mobile sources, where preemption is the general rule. The CAA preempts states from adopting any "standard relating to the control of emissions from new motor vehicles ..."[108] and the act defines "emission standard" as certain limits on "emissions of *air pollutants*."[109] The Supreme Court has now held that at least for purposes of mobile sources, "air pollutants" includes GHGs. Thus, CAA preemption of state regulation of car and truck GHG emissions is clear, whether or not EPA proceeds to regulate a particular mobile-source GHG. It would seem, then, that states are preempted from setting emission standards for CO_2, methane, and hydrofluorocarbons — three substances said to enhance climate change — even though EPA has not set mobile-source emission standards for them.

An exception to the general CAA rule preempting state mobile-source emission regulation is that EPA may waive CAA preemption for one particular state, California, if that state requests a waiver.[110] Further, when EPA does grant California a waiver, the act automatically extends it to almost all states with mobile-source emission limits identical to California's.[111]

Under this "California waiver" authority, California requested a preemption waiver for its GHG emissions regulations on December 21, 2005.

These regulations had been promulgated under a 2002 California enactment that was the first in the nation to call for limits on GHG emissions from mobile sources. Assembly Bill 1493[112] instructs the California Air Resources Board (CARB) to adopt regulations that achieve the maximum feasible reduction of GHGs emitted by passenger vehicles and light-duty trucks. The CARB adopted the required regulations in 2004. The regulations target CO_2, methane, nitrous oxide, and hydrofluorocarbon emissions, setting "fleet average greenhouse gas exhaust mass emission requirements for passenger car, light-duty truck, and medium-duty passenger vehicle weight classes." The first model year to which the fleet averages apply is 2009. The averages are reduced for each subsequent model year through 2016.

On December 19, 2007, almost two years after California requested the waiver, the EPA Administrator wrote the California governor that he intended to deny the state's request. On January 3, 2008, two petitions for review of this letter, arguing that it constituted final agency action on the waiver request, were filed in the Ninth Circuit. One suit was filed by California, with 18 other states intervening on California's side. The other suit was filed by environmental groups, and was consolidated with the first. However, since then, on February 29, 2008, the EPA Administrator signed a decision document formally announcing the waiver denial,[113] rendering unclear what will happen to the January 3 filings, which, again, are based on an argument that the *letter* is the reviewable agency action.[114] Presumably a new petition for review will be filed by the same or similar parties in connection with the formal decision document.[115]

Mobile Sources of GHG Emissions: Non-CAA Preemption

That the CAA preempts state GHG regulation of mobile sources cannot be seriously questioned, absent a California waiver. The following preemption litigation is significant for the *non-CAA* preemption claims being pressed. If successful, these claims would prevent California and other states from implementing the California mobile-source standards *even if EPA's denial of the waiver is judicially reversed.*

Non-CAA preemption suits, brought by auto interests, are pending in four of the federal judicial circuits containing a state that has adopted GHG controls on vehicles. Two decisions have been handed down, from Vermont (First Circuit) and California (Ninth Circuit). Both reject the non-CAA preemption theories presented.

In the first to be decided, *Green Mountain Chrysler Plymouth Dodge v. Crombie*, 508 F. Supp. 2d 295 (D. Vt. 2007), the court ruled that the relationship between Vermont's (California-identical) GHG standards and EPCA was better analyzed as an interplay between two federal statutes, rather than as a federal preemption question. So viewed, the court pointed out that the National Highway Traffic Safety Administration (NHTSA) has consistently treated EPA-approved California mobile source emissions standards as constituting "other motor vehicle standards of the Government," which EPCA says NHTSA must consider when setting CAFE standards.[116] Moreover, in a related context the *Massachusetts v. EPA* decision saw the EPCA CAFE provisions as harmonious with the CAA.[117] Thus, the court found the relationship between the CAA waiver authority and the EPCA CAFE provisions to be one of overlap, but not conflict. Despite its conclusion that preemption doctrine did not apply, the court did a preemption analysis anyway, finding that Vermont's GHG standards were preempted neither by EPCA nor as an intrusion upon the foreign policy authority of the United States.[118]

In the second decision, *Central Valley Chrysler-Jeep, Inc. v. Goldstene*, 529 F. Supp. 2d 1151 (E.D. Cal. 2007), a district court similarly rejected claims that California's first-in-the-nation GHG emission regulations were precluded by EPCA, preempted by EPCA, or preempted as an intrusion of state law on federal authority to conduct foreign affairs.

The legal theories pressed in the *Crombie* and *Goldstene* litigation are replicated in two Rhode Island suits — *Lincoln Dodge, Inc. v. Sullivan*, No. 1:06-CV-00070 (D.R.I. filed February 13, 2006), and *Association of International Automobile Manufacturers v. Sullivan*, No. 1:06-CV-00069 (D.R.I. filed February 13, 2006) — challenging that state's adoption of the California standards. Most recently, New Mexico's adoption of the California GHG standards has been challenged as preempted under EPCA in *Zangara Dodge, Inc. v. Curry*, No. 1 :07-CV-01305 (D.N.M. filed December 27, 2007).

VII. State Utility Laws

In 2000, the City of Seattle adopted a goal of meeting its electricity needs with "no net greenhouse gas emissions." To achieve this goal, the city ordered the city-owned electric utility to offset its GHG emissions by paying others to reduce their GHG emissions. The utility did so, largely through agreements paying other entities to use cleaner fuels. This made the utility (according to its press release) "the first large electric utility in the country to effectively eliminate its contribution of harmful greenhouse gas emissions." In *Okeson v. City of Seattle*, 150 P.3d 556 (Wash. 2007) (en banc), however, the utility's ratepayers argued that this offset arrangement was not authorized by the state's utility enabling act. The Washington Supreme Court agreed, explaining that the purchase of GHG offsets was not impliedly authorized by the enabling act in that the offset contracts were not proprietary because they were not part of the services for which ratepayers are billed, nor were they within the enabling act's purposes.

In another case, *In re Otter Tail Power Co.*, 744 N.W.2d 594 (S.D. 2008), environmental intervenors urged the state public utilities commission to deny a permit to build a coal-fired energy conversion facility, in light of the substantial CO_2 it would emit. Notwithstanding, the commission granted the permit, and the state supreme court sustained. The commission, it held, was not clearly erroneous in finding that the added CO_2 threatened no "serious" injury to the environment, the state's statutory standard. Deference to the commission was particularly apprpriate, it said, because the CO_2 from the plant would increase national CO_2 emissions by only .07%, and neither Congress nor the state had chosen to regulate CO_2 emissions.

VIII. INTERNATIONAL LAW

Could the United States, as a major GHG emitter that has declined to participate in the Kyoto Protocol, be sued in international forums for the adverse effects of climate change? Gauging the viability of such claims against the United States involves a good deal of guesswork, as they lie on the frontiers of international law. This report, concerned primarily with actually filed claims, notes only a few highlights, taken mostly from what appears to be the most pertinent article in the area.[119] The article suggests that the International Court of Justice (ICJ) might be one forum for resolution of climate change claims, with jurisdiction established through treaties that specifically provide for dispute resolution before the court. A problem with the ICJ approach is that the treaties most likely to be invoked are Friendship, Commerce, and Navigation or similar treaties, which focus on how each party *within its own country* treats the other country's nationals and property. A climate change suit, by contrast, likely would have an extraterritorial focus.

Another ICJ possibility would be for the court to render an advisory opinion, at the request of a body authorized under the U.N. Charter to request one.

Other possibilities include voluntary submission of a climate change dispute to any of several international arbitral forums or resort to the specialized dispute resolution systems created under various treaties. An example of the latter, reportedly being actively considered, is a fisheries conservation agreement under the UN Law of the Sea Convention, presumably on the argument that increased ocean temperatures from climate change imperil certain fish stocks.[120]

Some principles that might be applied to a claim alleging GHG-caused injury might be taken from the international law of transboundary pollution. For example, the Restatement (Third) of Foreign Relations Law describes an international law principle under which a nation must "take such measures as may be necessary, to the extent practicable under the circumstances, to ensure that activities within its jurisdiction or control ... are conducted so as not to cause significant injury to the environment of another state...."[121] Similarly, the *Trail Smelter* arbitration decision, probably the seminal ruling on state liability for transboundary pollution, declared that "[a] State owes at all times a duty to protect other States against injurious acts by individuals from within its jurisdiction."[122] Of course, as with the domestic common law litigation described in Section V, daunting hurdles confront the claimant in making the link between climate change in general and specific environmental harms, and in apportioning how much of such harms are attributable to the charged party or parties, in this instance the United States.

A widely noted international-law action was filed by a group based in the Arctic, where the temperature rise from climate change has been among the fastest. In 2005, the Chair of the Inuit Circumpolar Conference, on behalf of herself and all affected Inuit of the arctic regions of the United States and Canada, filed a petition against the United States with the Inter-American Commission on Human Rights, the investigative arm of the Organization of American States (OAS).[123] The petition alleged that the United States, through its failure to restrict its GHG emissions and the resultant climate change, has violated the Inuit's human rights — including their rights to their culture, to property, to the preservation of health, life, and physical integrity, and so on.[124] Inuit culture is described in the petition as "inseparable from the condition of [its] physical surroundings." Generally, the Inter-American Commission on Human Rights is empowered to recommend measures that contribute to human rights protection, request states in urgent cases to adopt specific precautionary measures to avoid serious harm to human rights, or submit cases to the Inter-American Court of Human Rights. The United States, however, has not accepted the jurisdiction of this court, so the Inuit petition sought only to have the Commission prepare a report declaring the responsibilities of the United States and recommending corrective measures.

In 2006, the Inuit petition was rejected, with no reasons given (as is customary for the Commission). However, at the request of petitioners the Commission held a hearing on March 1, 2007 on the generic issue of climate change and human rights. One may speculate that the Commission felt more comfortable with the hearing format than the petition because the former did not single out the United States. Or that the Commission was concerned the petition took it into a realm of global scale, orders of magnitude vaster than the typical human rights petition it receives.

In 2005/2006, five petitions were submitted to the Intergovernmental Committee for the Protection of the Cultural and Natural Heritage of Outstanding Universal Value (World Heritage Committee), part of UNESCO.[125] The petitions request that various designated World Heritage Sites be placed on the List of World Heritage in Danger[126] owing to alleged impacts of climate change. The sites covered by the petitions are Waterton-Glacier International Peace Park (U.S./Canada), Sagarmatha National Park (Nepal), Belize Barrier Reef Reserve System (Belize), Huascaran National Park (Peru), and the Great Barrier Reef (Australia). Only the Waterton-Glacier petition was filed by entities within the United States (12 environmental groups) and involves a natural resource within the United States. As a party to the World Heritage Convention, the United States is obligated to "do all it can ... to the utmost of its own resources and, where appropriate, with any international assistance and cooperation" to protect its cultural and natural heritage.[127]

In 2006, the World Heritage Committee acknowledged the five petitions but appeared desirous of shifting the debate toward the use of existing committee mechanisms at individual sites to adapt to the threat of climate change.[128]

IX. COMMENTS

Gauging the prospects of the pending climate change lawsuits is a precarious venture; for many of the suits, there is little precedent. It is clear, however, that success in the conventional sense — obtaining a judgment for the environmental plaintiff — is not the only objective of many of these suits. Some of the climate change litigation almost certainly has a long-range strategic purpose — to keep climate change on the political front burner and make it difficult for government and GHG emitters to ignore the problem.

In the conventional sense of the term, plaintiffs' successes have been rare in cases seeking relief *directly from GHG emitters*. A court may be reluctant to impose expensive measures to address a global problem on a defendant that is a proportionately minor contributor (which almost all defendants are, given the vast number of GHG emitters), using statutory provisions or common law principles that were not formulated with global problems in mind, against a backdrop of scientific uncertainty as to the precise consequences (if not the general cause) of climate change. By contrast, the environmental side recently has scored major wins where *governmental* remedies were sought. In a much-publicized string of 2007 decisions under the Clean Air Act,[129] Energy Policy and Conservation Act of 1975,[130] foreign policy authority of the United States,[131] and NEPA,[132] courts have shown increased willingness to authorize or require government consideration of climate change.

As this report shows, standing has been a persistent issue for environmental plaintiffs, though of late the tide appears to be shifting in their favor. And at least for states, the Supreme Court decision in *Massachusetts v. EPA* is likely to work a sea change in improving plaintiffs' prospects. As noted earlier, the big question is the extent to which the Supreme Court decision finding standing will be seen by the lower courts to generalize to nonstate plaintiffs, other statutory and common law contexts, and injuries (as from weather events) not as clearly attributable to climate change as Massachusetts's loss of shore land.

Causation is not only a component of the threshold standing test but a component of the plaintiff's case on the merits. Several writers have identified proof of causation as a key obstacle to a tort action seeking relief from climate change injury.[133] And at the remedy stage, allocation of damages among specific defendants will likely present both factual and legal challenges.

In either the standing or case-in-chief contexts, the climate change issues in private-remedy actions reprise an intractable problem in environmental law: imposing liability for harms that are remote in time and place from the pollution sought to be abated, particularly where the pollution comes from multiple sources.[134] Lawmakers of yesteryear encountered this same redistributive conundrum in tackling the problem of acid rain, where pollution cause and effect are separated by hundreds of miles and weeks or months. Imposing liability for harm from exposure to toxic chemicals is of the same ilk: exposure to contamination from multiple sources may result in ill effects manifested only a decade or two later.

Perhaps because of these hurdles under existing law, and the resistance of the current Administration to regulatory approaches to climate change, new directions are now being explored.[135] Within the United States, several states have adopted their own GHG emission controls, citing, among other things, inaction at the federal level.[136] Twenty-three states have joined one of the three regional GHG reduction initiatives (Western Climate Initiative, Midwestern Regional Greenhouse Gas Reduction Accord, and in the eastern states, Regional Greenhouse Gas Initiative). Some states have explored the idea of emissions trading with Europe.[137] Use of the shareholder proposal process and SEC disclosure requirements have been suggested as ways of forcing the issue in the corporate world.[138] More loosely, 31 states and two Canadian provinces announced in May 2007 the formation of a Climate Registry to track greenhouse gas emissions from major industries, with independent verification of data. California and the United Kingdom signed an agreement on July 31, 2006, committing both parties to implement market-based mechanisms, to share results from studies to quantify the economic impacts of climate change, collaborate on research, etc.[139] Also internationally, this report noted the unsuccessful Inuit petition filed with the Inter-American Commission on Human Rights and the pending petitions before the World Heritage Committee. Reportedly, the low-lying Pacific nation of Tuvalu threatened to sue the United States and Australia four years ago in the ICJ, but held off for unspecified reasons.[140]

Whether these new paths will yield results, only time will tell. It is clear, however, that if there is to be a government response to climate change at all, a solution from the political branches is more likely to be comprehensive and fully reflective of societal priorities than the typically narrowly targeted results of litigation. Many proponents of litigation or unilateral state action freely concede that such initiatives are make-do efforts that, while making a small contribution to mitigating climate change, are also aimed at prodding the national government to act.

End Notes

[1] 127 S. Ct. 1438 (2007).

[2] Similar ground is covered by Justin R. Pidot, GLOBAL WARMING IN THE COURTS: AN OVERVIEW OF CURRENT LITIGATION AND COMMON LEGAL ISSUES (Georgetown Environmental Law and Policy Institute 2006) (available, together with a March, 2007

update, at [http://www.law and Todd O. Madden and Eric McLaughlin, *Climate Change Litigation: Trends and Developments*, BNA Daily Env't Rpt. B-1 (April 3, 2007). A regularly updated chart of climate change cases, prepared by Michael Gerrard of Arnold & Porter, is available at [http://www.abanet.org/abapubs/globalclimate/]. Broader treatments of the legal implications of climate change may be found in Michael Gerrard (ed.), GLOBAL CLIMATE CHANGE AND U.S. LAW (ABA 2007), and Symposium Issue, RESPONSES TO GLOBAL WARMING: THE LAW, ECONOMICS, AND SCIENCE OF CLIMATE CHANGE, 155 U. Pa. L. Rev. 1353 et seq. (2007).

[3] An example of a case that deals with climate change only implicitly (at least so far) is *State of New York v. U.S. Dep 't of Energy*, No. 08- 0311 (2d Cir. filed January 17, 2008), in which three states (NY, CN, MA) challenge the Department's energy conservation standards for residential furnaces and boilers. Though we are given to understand that the climate change benefits of reducing fossil fuel consumption by such furnaces and boilers was a consideration in filing suit, the petition for review does not mention CO_2 or climate change, and thus we do not include this case in the body of the report.

[4] Kyoto Protocol to the United Nations Framework Convention on Climate Change, concluded December 10, 1997, U.N. Doc. FCC/CP/1997/L.7 Add. 1, reprinted at 37 I.L.M. 22 (1998). One indication of Senate antipathy to the Kyoto Protocol was its adoption by 95- 0 of the so-called Byrd-Hagel resolution urging the President not to sign any international agreement on climate change that would result in serious injury to the U.S. economy or that did not include provisions regarding the GHG emissions of developing countries. S. 98, 105th Congress (1997).

[5] *See generally* CRS Report 98-349, *Global Climate Change: Selected Legal Questions About the Kyoto Protocol*, David M. Ackerman. This report concluded that "there does not appear to be any clear legal authority that could be invoked to sustain the provisional application of the Kyoto Protocol." *Id.* at 6.

[6] P.L. 105-276, 112 Stat. at 2496 (1998) (barring EPA's use of FY1999 funds to implement Protocol); P.L. 106-74, 113 Stat. at 1080 (1999) (same for FY2000); P.L. 106-377, 114 Stat. at 1 141A-41 (2000) (same for FY2001).

[7] Memorandum from Jonathan Z. Cannon, EPA General Counsel, to Carol M. Browner, EPA Administrator, EPA's Authority to Regulate Pollutants Emitted by Electric Power Generation Sources (April 10, 1998).

[8] *Is CO_2 A Pollutant and Does EPA Have the Power to Regulate It?*, Joint Hearing Before the Subcomm. on National Environmental Growth, Natural Resources and Regulatory Affairs of the House Comm. on Gov't Reform and the Subcomm. on Energy and Environment of the House Comm. on Science, 106th Cong. (1999).

[9] Testimony of Gary Guzy, Joint Hearing, *supra* note 8, at 11.

[10] See Veronique Bugnion and David M. Reiner, *A Game of Climate Chicken: Can EPA Regulate Greenhouse Gases Before the U.S. Senate Ratifies the Kyoto Protocol?*, 30 Envtl. L. 491 (2000).

[11] Civ. Action No. 3:03CV984 (PCD) (D. Conn.) (filed June 4, 2003).

[12] 42 U.S.C. § 7408.

[13] 42 U.S.C. § 7409.

[14] 71 Fed. Reg. 9866 (February 27, 2006).

[15] Coke Oven Environmental Task Force v. EPA, No. 06-1131 (D.C. Cir. filed April 7, 2006).

[16] No. 06-1322.

[17] 434 F. Supp. 2d 957 (D. Or. 2006).

[18] CAA § 165, 42 U.S.C. § 7475.

[19] CAA § 165(a)(4), 42 U.S.C. § 7475(a)(4); CAA § 169(3), 42 U.S.C. § 7479(3).

[20] 42 U.S.C. § 7521(a)(1).

[21] EPA, Control of Emissions from New Highway Vehicles and Engines, 68 Fed. Reg. 52922 (September 8, 2003). EPA's ruling followed a suit by the original petitioners alleging

unreasonable delay. Center for Technology Assessment v. Whitman, No. 02-CV-2376 (D.D.C. filed December 5, 2002).

[22] Memorandum from Robert E. Fabricant, EPA General Counsel, to Marianne L. Horinko, EPA Acting Administrator, EPA's Authority to Impose Mandatory Controls to Address Global Climate Change Under the Clean Air Act (August 28, 2003).

[23] 415 F.3d 50 (D.C. Cir. 2005).

[24] Emphasis added.

[25] The test for whether a plaintiff in federal court has standing is described in greater detail in the following discussion of the Supreme Court's decision on appeal of the D.C. Circuit ruling.

[26] 127 S. Ct. 1438 (2007).

[27] *Id.* at 1446.

[28] CAA § 307(b)(1), 42 U.S.C. § 7607(b)(1).

[29] 127 S. Ct. at 1453.

[30] *Id.* at 1454.

[31] *Id.*

[32] *Id.* at 1457.

[33] *Id.* at 1458.

[34] 49 U.S.C. § 32902.

[35] 127 S. Ct. at 1463.

[36] Three weeks after the decision in *Massachusetts v. EPA*, the Senate held a hearing devoted exclusively to it: *The Implications of the Supreme Court's Decision Regarding EPA's Authorities with Respect to Greenhouse Gases Under the Clean Air Act: Hearing Before the Senate Comm. on Env 't and Pub. Works* (April 24, 2007) (hereinafter *Senate hearing*).

[37] Justice Scalia's dissent characterizes EPA's three options similarly: 127 S. Ct. at 1472.

[38] *Senate hearing, supra* note 36 (prepared statement of EPA Administrator Stephen Johnson). The EPA Administrator is apparently referring to the Court's statement that "We need not and do not reach the question ... whether policy concerns can inform EPA's actions in the event that it makes [an endangerment finding]." 127 S. Ct. at 1463.

[39] *Briefing by Conference Call on the President's Announcement on CAFE and Alternative Fuel Standards*, May 14, 2007 (statement of EPA Administrator Stephen Johnson), available at [whitehouse.gov/news/releases/2007]. *See also* Exec. Order No. 13432, 2007 Westlaw 1405388 (White House), signed May 14, 2007.

[40] Pub. Law 110-140.

[41] *See* letters to EPA Administrator Stephen Johnson from 17 states and from several environmental groups, both dated January 23, 2008, charging that EPA has unreasonably delayed in responding to the *Massachusetts v. EPA* remand. Both letters charge that the enactment of the Energy Independence and Security Act provides no basis for not moving forward on the remand.

[42] Letter from EPA Administrator Stephen Johnson to Senators Barbara Boxer and James Inhofe, March 27, 2008; identical letter from Administrator Johnson to Representatives John Dingell and Joe Barton, March 27, 2008.

[43] Petition for writ of mandamus to compel compliance with mandate, filed April 2, 2008. The petition argues that the ANPR, a broad request for comments on the role of GHGs throughout the CAA, is similar to one of the policy arguments made by EPA in denying the original 1999 petition, namely that establishing motor vehicle GHG standards now —

would result in an inefficient piecemeal approach to addressing the climate change issue.... A sensible regulatory scheme would require that all significant sources and sinks of GHG emissions be considered in deciding how best to achieve any needed emission reductions.

68 Fed. Reg. at 52,391. As noted, the Supreme Court decision held that such policy considerations are irrelevant to the agency's determination on the section 202 petition.

[44] *See, e.g.*, CAA § 108(a)(1)-(2), 42 U.S.C. § 7408(a)(1)-(2) (requiring the EPA Administrator to maintain a list of each "air pollutant" "emissions of which, in his judgment, cause or contribute to air pollution which may reasonably be anticipated to endanger public health or

welfare," and then issue air quality criteria and national ambient air quality standards for that air pollutant).

[45] CAA § 165, 42 U.S.C. § 7475.

[46] CAA § 165(a)(4), 42 U.S.C. § 7475(a)(4).

[47] 42 U.S.C. § 7547(a)(4). The petition recognizes that while CAA section 202(a)(1) at issue in *Massachusetts v. EPA* states that EPA "shall" issue standards upon making an endangerment finding, section 213(a)(4) states that EPA "may" do so.

[48] 42 U.S.C. § 7571.

[49] See Sara Zdeb, *From* Georgia v. Tennessee Copper *to* Massachusetts v. EPA: Parens Patriae *Standing for State Global-Warming Plaintiffs*, 96 Geo. L. J. 1059 (2008).

[50] California v. General Motors Corp., 2007 Westlaw 2726871 (N.D. Cal. September 17, 2007). This case is discussed in Section V.

[51] 16 U.S.C. §§ 1361-1421h.

[52] 16 U.S.C. § 137 1(a)(5). In the MMPA, "take" means "to harass, hunt, capture, or kill" any marine mammal, or attempt to do so. 16 U.S.C. § 1362(13).

[53] 71 Fed. Reg. 43,926 (August 2, 2006).

[54] In a case of the same name, *Center for Biological Diversity v. Kempthorne*, No. 07-5109 (N.D. Cal. filed October 4, 2007), environmental groups challenge the Secretary of the Interior's alleged failure to issue updated stock assessment reports for marine mammals under its jurisdiction (sea otters, polar bears, walrus, and manatees) within the time frames mandated by the MMPA. The complaint asserts as examples that since the last stock assessment reports on the polar bear and walrus, "global warming has caused the loss of sea ice upon which [those species] depend...."

[55] 16 U.S.C. §§ 1531-1544. The ESA defines "take" similarly to the MMPA, note 52. 16 U.S.C. § 1532(19).

[56] 16 U.S.C. § 1538(a)(1)(B)-(C).

[57] By general rule, the Fish and Wildlife Service has extended all of the endangered species prohibitions to threatened animals. 50 C.F.R. § 17.31. "Special rules," withdrawing particular threatened species from aspects of the general regime, have been promulgated for those species with atypical management needs, such as grizzly bears. 50 C.F.R. § 17.40(b).

[58] 16 U.S.C. § 1536(a)(2). Because section 7 is more easily triggered when the species' habitat receives a formal designation as "critical habitat," litigation to compel such designation is another aspect of environmental groups' efforts to use the ESA against global warming. *See* ESA § 4(a)(3), 16 U.S.C. § 1533(a)(3).

[59] 89 Fed. Reg. 26852 (May 9, 2006). The Center for Biological Diversity has also settled a suit requiring NMFS to designate critical habitat for ESA-listed corals. The proposed critical habitat rule is at 73 Fed. Reg. 6895 (February 6, 2008).

[60] 72 Fed. Reg. 1,064 (January 9, 2007). Under ESA section 4(b)(6)(A), the Secretary of the Interior is required to make his determination for or against listing within one year of the proposed listing — unless a six-month extension is claimed under ESA section 4(b)(6)(B). No extension having been claimed, the polar bear determination was due by January 9, 2008, but had not been made as of the date of this report. On March 10, 2008, environmental groups sued the Secretary of the Interior to compel an up or down determination. *Center for Biological Diversity v. Kempthorne*, No. 08-1339 (N.D. Cal. filed March 10, 2008).

[61] *See* CRS Report RL3394 1, *Polar Bears: Proposed Listing Under the Endangered Species Act*, by Eugene H. Buck.

[62] The Center for Biological Diversity has filed suit over the FWS's delay in making its "12-month finding" regarding 10 species of penguin. *Center for Biological Diversity v. Hall*, No. 1:08-cv-00335 (D.D.C. filed February 27, 2008). Under ESA section 4(b)(3)(B), the Secretary of the Interior has 12 months after receiving a petition indicating that action may be warranted to either propose listing or find that no action is warranted.

[63] *See generally* Brendan R. Cummings and Kassie R. Siegel, Ursus maritimus: *Polar Bears on Thin Ice*, Natural Res. & Env't (ABA) 3 (Fall 2007) (discussing how "the listing process for the polar bear highlights the possibilities and limitations of using the ESA to address otherwise unregulated GHG emissions").

[64] 50 C.F.R. §§ 17.3 (Fish and Wildlife Service), 222.102 (NOAA Fisheries).

[65] 49 U.S.C. § 32902(a).

[66] 2007 Westlaw 3378240, *16.

[67] 43 U.S.C. § 1331 et seq.

[68] Taken from Petitioner Center for Biological Diversity's Non-Binding Statement of Issues, filed August 3, 2007.

[69] An apparently exhaustive survey of the NEPA/climate change cases, decided and pending, is Joseph Mendelson III (Legal Director, Center for Food Safety and International Center for Technology Assessment), *Surveying the National Environmental Policy Act and the Emerging Issues of Climate Change, Genetic Engineering and Nanotechnology* (October 30, 2007) (copy on file with author).

[70] *See also* International Center for Technology Assessment et al., *Petition requesting that the Council on Environmental Quality amend its regulations to clarify that climate change analyses be included in environmental review documents* (filed February 28, 2008).

[71] Other model years were involved, too, but only the challenge to the model year 1989 CAFE standard involved a climate change argument.

[72] 912 F.2d at 501.

[73] Florida Audubon Society v. Bentsen, 94 F.3d 658 (D.C. Cir. 1996). The four dissenting judges argued that the majority had "misapplied the doctrine of standing to the assertion of a procedural right, such as the preparation of an EIS, with the consequence that it will be effectively impossible for anyone to bring a NEPA claim in the context of a rulemaking with diffuse impact." *Id.* at 673.

[74] *See, e.g.*, Friends of the Earth v. Laidlaw Environmental Services, 528 U.S. 167 (2000).

[75] 794 F.2d at 401.

[76] In finding standing, the judge repudiated an earlier climate change/standing decision of the same court. In *Center for Biological Diversity v. Abraham*, 218 F. Supp. 2d 1143 (N.D. Cal. 2002), plaintiffs had sought enforcement of the Energy Policy Act as it related to the acquisition of alternative fuel vehicles by the United States. In rejecting standing, this decision spurned plaintiffs' climate change concerns as "too general, too unsubstantiated, too unlikely to be caused by defendant's conduct, and/or too unlikely to be redressed by the relief sought to confer standing." In *Friends of the Earth*, the court neutralized this pronouncement by noting that "*Center for Biological Diversity* was decided before the Ninth Circuit clarified in [Citizens for Better Forestry v. U.S. Dep't of Agriculture, 341 F.3d 961, 972 (9th Cir. 2003)] that environmental plaintiffs raising procedural concerns need not present proof that the challenged federal project will have particular environmental effects."

[77] An organization has standing to bring suit on behalf of its members when "(a) its members would otherwise have standing to sue in their own right; (b) the interests it seeks to protect are germane to the organization's purpose; and (c) neither the claim asserted nor the relief requested requires the participation of individual members in the lawsuit." Hunt v. Washington State Apple Advertising Comm'n, 432 U.S. 333, 343 (1977).

[78] 260 F. Supp. 2d at 1028.

[79] 71 Fed. Reg. 17, 566 (April 6, 2006).

[80] 508 F.3d at 550.

[81] *Id.*

[82] 71 Fed. Reg. 43,926 (August 2, 2006).

[83] 40 C.F.R. § 1508.8.

[84] *See* 345 F.3d at 550.

[85] *See* Dave Owen, *Climate Change and Environmental Assessment Law*, 33 Colum. J. Env'l L. 57 (2008).

[86] The attorney general lawsuit is State of California v. County of San Bernardino, No. CIVSS07-00329 (Cal. Super. Ct. filed April 12, 2007).

[87] 15 U.S.C. §§ 2921-2961.

[88] *See* 15 U.S.C. § 2931(a) (congressional findings suggestive of the act's scope).

[89] 15 U.S.C. § 2934(a).

[90] 15 U.S.C. § 2936.

[91] The summary was published at 72 Fed. Reg. 73771 (December 28, 2007).

[92] 5 U.S.C. § 552.

[93] Revised Joint Briefing Schedule Statement, dated August 1, 2007.

[94] Renamed the House Committee on Oversight and Government Reform early in the 110th Congress.

[95] This nutshell on possible plaintiffs and defendants is adapted from David Hunter and James Salzman, *Negligence in the Air: The Duty of Care in Climate Change Litigation*, 155 U. Pa. L. Rev. 1741, 1750-1752 (2007).

[96] American Electric Power Co., Inc., The Southern Co., Cinergy Corp., Tennessee Valley Authority, and Xcel Energy, Inc.

[97] An activity is a public nuisance if it creates an unreasonable interference with a right common to the general public. Unreasonableness may rest on the fact that the activity significantly interferes with public health and safety, or has produced a permanent or long- lasting effect and, as the actor knows or has reason to know, has a significant effect on the public right. Restatement (Second) of Torts § 821B (1979).

[98] An activity is a private nuisance if it is a nontrespassory invasion of another's interest in the private use and enjoyment of land. *Id.* at § 821D.

[99] *See* Vincent S. Oleskiewicz and Douglas B. Sanders, *The Advent of Climate Change Litigation Against Corporate Defendants*, BNA Daily Env't Rpt. B-1 (November 15, 2004). The authors review the *State of Connecticut* and *Open Space Institute* suits in some detail, assess the defenses available in tort-based climate change suits generally, and extract clues as to the potential success of such litigation from the history of litigation against tobacco companies.

[100] *See generally* Thomas W. Merrill, *Global Warming as a Public Nuisance*, 30 Colum. J. Envtl. L. 293 (2005) (discussing threshold hurdles of standing, the applicability of federal common law, and foreign policy preemption).

[101] An interesting question raised by the Prof. Merrill article, *supra* note 100, is whether these general standing requirements, developed in the context of private lawsuits, should apply in a suit such as *State of Connecticut* — that is, in a parens patriae suit brought by state attorneys general under public nuisance law.

[102] Connecticut v. American Elec. Power, 406 F. Supp. 2d 265 (S.D.N.Y. 2005).

[103] Baker v. Carr, 369 U.S. 186, 210 (1962).

[104] *Id.* at 217.

[105] Presumably two of the plaintiffs, New York State and New York City, have been able to support their standing to sue by arguments not contrary to those they made *against* plaintiff standing in another public-nuisance climate change case in which they were the defendants. Avoiding contradictory arguments was presumably facilitated by the idiosyncratic nature of the global-warming harms alleged by the *pro se* plaintiff — e.g., those based on plaintiff's enhanced vulnerability to disease-causing pollution as compared to the general population. According to the court, plaintiff appeared to be requesting an injunction ordering the defendants to stop polluting and use his invention for reducing carbon dioxide emissions. Korsinsky v. U.S. EPA, 192 Fed. Appx. 171 (2d Cir. 2006) (affirming district court dismissal based on lack of standing).

[106] CAA § 116, 42 U.S.C. § 7416. The exceptions in this nonpreemption provision say that states may not adopt emission limitations for stationary sources that are *less* stringent than those in

state implementation plans, new source performance standards, or national emission standards for hazardous air pollutants.

[107] *See* Global Warming Solutions Act of 2006, A.B. 32, Cal. Health & Safety Code § 38500 et seq. This law requires that GHG emission limits be in effect in California by 2012 to reduce statewide GHG emissions to the 1990 level by 2020. Note, however, that although A.B. 32 applies chiefly to stationary sources, it provides that if the mobile source GHG emission limits imposed by an earlier state enactment are struck down, "alternative regulations" to restrict mobile-source GHG emissions shall be imposed under A.B. 32. As the following paragraphs of the text discuss, this earlier enactment is now the subject of a preemption challenge.

[108] CAA § 209(a), 42 U.S.C. § 7543(a).

[109] CAA § 302(k), 42 U.S.C. § 7602(k). Emphasis added.

[110] CAA § 209(b), 42 U.S.C. § 7543(b). Under section 209(b), EPA "shall" grant the waiver "if the State determines that the State standards will be, in the aggregate, at least as protective of public health and welfare as applicable federal standards." However, no waiver shall be granted if EPA finds that the state's determination is arbitrary and capricious, the state does not need the standards to meet "compelling and extraordinary conditions," or the state standards and accompanying enforcement procedures are inconsistent with CAA section 202(a).

[111] CAA § 177, 42 U.S.C. § 7507. Section 177 limits its extension of the section 209 waiver to those states that have approved nonattainment-area plans. This includes all states except North Dakota, South Dakota, and Hawaii.

[112] Cal. Heath & Safety Code § 43018.5.

[113] 73 Fed. Reg. 12156 (March 6, 2008).

[114] It is *not* unclear, however, what will happen to California's November, 2007, suit asking the court to order EPA to issue its waiver decision forthwith. This suit is plainly moot now.

[115] *See generally California's Waiver Request to Control Greenhouse Gases Under the Clean Air Act*, CRS Report RL34099, by James E. McCarthy and Robert Meltz.

[116] 49 U.S.C. § 32902(f).

[117] 127 S. Ct. at 1462.

[118] *See, e.g.*, American Ins. Ass'n v. Garamendi, 539 U.S. 396 (2003).

[119] Andrew L. Strauss, *The Legal Option: Suing the United States in International Forums for Global Warming Emissions*, 33 Envtl. L. Rptr. 10185 (2003).

[120] Agreement for the Implementation of the Provisions of the United Nations Convention on the Law of the Sea of 10 December 1982 Relating to the Conservation and Management of Straddling Fish Stocks and Highly Migratory Fish Stocks, 34 Int'l Legal Materials 1547. The United States is a party to this agreement and, by reference to the Law of the Sea Convention, it incorporates binding dispute-resolution mechanisms.

[121] Restatement (Third) of Foreign Relations Law § 601(1). *See also* Legality of the Threat or Use of Nuclear Weapons, Advisory Opinion, 1996 ICJ Reports 226, 24 1-242 (July 8, 1996) ("the existence of the general obligation of states to ensure that activities within their jurisdiction and control respect the environment of other states or of areas beyond national control is now part of the corpus of international law relating to the environment").

[122] Trail Smelter (U.S. v. Canada), 3 R.I.A.A. 1938, 1965 (March 11, 1941).

[123] For an eight-page summary of the 176-page petition, go to [http://earthjustice.org/library/legal_docs/summary-of-inuit-petition-to-inter-american-council-on-human-rights.pdf].

[124] *See generally* Sara C. Aminzadeh, Note, *A Moral Imperative: The Human Rights Implications of Climate Change*, 30 Hastings Int'l & Comp. L. Rev. 231 (2007).

[125] Convention Concerning the Protection of the World Cultural and Natural Heritage, art. 8, signed November 23, 1972, entered into force December 17, 1975, 27 U.S.T. 37.

[126] *Id.* at art.11, par. 4.

[127] *Id.* at art. 4.

[128] World Heritage Comm. Decision 30 COM 7.1, adopted July 10, 2006, available at [http://law

[129] Massachusetts v. EPA, 127 S. Ct. 1438 (2007) (see Section I of this report).

[130] Green Mountain Chrysler Plymouth Dodge Jeep v. Crombie, 508 F. Supp. 2d 295 (D. Vt. 2007), and Central Valley Chrysler-Jeep, Inc. v. Goldstene, 529 F. Supp. 2d 1151 (E.D. Cal. 2007) (both described in section VI of this report); Center for Biological Diversity v. National Highway Traffic Safety Administration, 508 F.3d 508 (9th Cir. 2007) (described in sections III and IV of this report).

[131] *Green Mountain*, *supra* note 130.

[132] *Center for Biological Diversity*, *supra* note 130.

[133] Myles R. Allen and Richard Lord, *The Blame Game: Who Will Pay for the Damaging Consequences of Climate Change?*, 432 Nature 551 (December 2004); David A Grossman, *Warming Up to a Not-So-Radical Idea: Tort-Based Climate Change Litigation*, 28 Colum. J. Envtl. L. 1 (2003): Eduardo M. Penalver, *Acts of God or Toxic Torts? Applying Tort Principles to the Problem of Climate Change*, 38 Nat. Res. J. 563, 569 (1998).

[134] *See generally* Richard J. Lazarus, THE MAKING OF ENVIRONMENTAL LAW ch. 1 (2004).

[135] *See, e.g.*, Kristin Marburg, *Combating the Impacts of Global Warming: A Novel Legal Strategy*, 2001 Colo. J. Int'l L. & Pol'y 171 (2001).

[136] *See, e.g.*, California's A.B. 32, the Global Warming Solutions Act of 2006, *supra* note 107. *See generally* CRS Report RL338 12, *Climate Change: Action by States to Address Greenhouse Gas Emissions*, by Jonathan L. Ramseur.

[137] Congressional Green Sheets Newsroom, December 17, 2004. The same source reports that Rep. Joe Barton (R-Texas), then-chairman of the House Energy and Commerce Committee, said that any international compact involving state governments would have to be approved by Congress and that "we would tend to look at it with a lot of skepticism." Kenneth Colburn, who is helping to coordinate the states' effort, was said to question the need for federal authorization, on the theory that any transatlantic trades would be commercial transactions, not government-to-government.

[138] *See, e.g.*, Sung Ho (Danny) Choi, Note, *It's Getting Hot in Here: The SEC's Regulation of Climate Change Shareholder Proposals Under the Ordinary Business Exception*, 17 Duke Envtl. L. & Pol'y Forum 165 (2006); California Public Employees' Retirement System et al., *Petition for interpretive guidance on climate change disclosure*, SEC No. 4-547 (submitted September 18, 2007); Free Enterprise Action Fund, *Petition for interpretive guidance under the Securities Act of 1933 that would require registrants to disclose to shareholders the business risks of laws and regulations intended to address global warming concerns*, SEC No. 4-549 (submitted October 22, 2007).

[139] United Kingdom and California Announcement on Climate Change and Clean Energy Collaboration.

[140] *See* [http://www.tuvaluislands.com]. Tuvalu alleged that Australia is the biggest per capita producer of GHGs, and that the United States is the biggest single emitter. *See also* Aurelie Lopez, *The Protection of Environmentally Displaced Persons in International Law*, 37 Envtl. L. 365, 372-373 (2007). Residents of the Alaskan village of Shishmaref on the Bering Strait, who are now being relocated, are apparently the first American climate change refugees.

In: Climate Change Litigation and Law ISBN: 978-1-60876-089-3
Editors: Jean-François Masson

Chapter 2

California's Waiver Request under the Clean Air Act to Control Greenhouse Gases from Motor Vehicles

James E. McCarthy and Robert Meltz

Summary

California has adopted regulations requiring new motor vehicles to reduce emissions of greenhouse gases (GHGs), beginning in model year 2009. The Clean Air Act (CAA) generally preempts states from adopting their own emission standards for mobile sources. However, the act allows such standards in California, if the state obtains a waiver of CAA preemption from EPA.

California requested this waiver in 2005, but EPA took until December 19, 2007, to decide that it would deny the request. On that day, EPA Administrator Stephen Johnson wrote California Governor Schwarzenegger to say, "I have decided that EPA will be denying the waiver and have instructed my staff to draft appropriate documents setting forth the rationale for this denial in further detail.... " According to press reports, the decision was taken against the unanimous advice of the agency's technical and legal staffs. The

Administrator published a decision document denying the waiver in the March 6, 2008 *Federal Register*.

Following EPA's denial of the waiver request, California and environmental groups petitioned for review in the D.C. Circuit, with 18 other states intervening on California's side. The interest of the intervening states derives from the fact that under the CAA, other states may adopt motor vehicle emission standards identical to California's and avoid CAA preemption if California is granted a waiver. Fourteen states and the District of Columbia have already adopted such regulations.

This report reviews the nature of EPA's, California's, and other states' authority to regulate emissions from mobile sources, the applicability of that authority to GHGs, and issues related to the California waiver request. To obtain a waiver, California must meet conditions laid out in CAA Section 209(b): the state must first determine that its standards will in the aggregate be at least as protective of public health and welfare as applicable federal standards. The EPA Administrator must then find whether the state's determination is arbitrary and capricious; whether the state needs the standards to meet compelling and extraordinary conditions; and whether the standards and accompanying enforcement procedures are consistent with CAA Section 202(a).

This report does not analyze whether California is preempted from regulating mobile-source GHGs by the Corporate Average Fuel Economy (CAFE) requirements of the Energy Policy and Conservation Act of 1975, or the amended CAFE standards of the Energy Independence and Security Act of 2007 (P.L. 110-140). Under these laws, authority to set fuel economy standards is reserved to the federal government—specifically, the National Highway Traffic Safety Administration (NHTSA). In several court cases and in other venues, the auto industry is maintaining that the regulation of mobile-source GHG emissions is simply another method of regulating fuel economy, so California's GHG standards (and identical standards adopted by other states) are preempted. Two federal district courts have rejected this argument, but both decisions have been appealed.

On January 21, 2009, California submitted a formal request to President Obama and EPA Administrator-designate Lisa Jackson for reconsideration of the waiver denial. In response to this request, EPA will hold a hearing March 5, 2009, and will accept written comments until April 6.

INTRODUCTION

Every federal law imposing environmental standards raises the question of whether the states are allowed to set stricter standards. In deference to states' rights, Congress's usual approach is to allow stricter state standards; for example, the Clean Air Act (CAA) allows stricter state standards for stationary sources of air pollution (power plants, refineries, etc.). For mobile sources of air pollution, however—cars, trucks, planes, etc.—a lack of national uniformity creates a problem, since manufacturers would potentially face the task of complying with different standards in each state. Such standards would fragment the national market, increasing costs and complicating the manufacture, sale, and servicing of the affected products. For this reason, the mobile source portion of the CAA (Title II) generally "preempts" states from adopting their own emission standards for new motor vehicles or engines.[1] In general, it allows only federal standards for motor vehicle emissions.

There is an exception to this rule of federal preemption, however, in CAA Section 209(b)[2]—

> The [EPA] Administrator shall, after notice and opportunity for public hearing, waive application of this section [the preemption of State emission standards] to any State which has adopted standards (other than crankcase emission standards) for the control of emissions from new motor vehicles or new motor vehicle engines prior to March 30, 1966, if the State determines that the State standards will be, in the aggregate, at least as protective of public health and welfare as applicable Federal standards.[3]

Only California adopted such standards "prior to March 30, 1966," so only California can qualify for such a preemption waiver. (Although only California may be granted a waiver under this section, elsewhere in the Act, as discussed later in this report, there is a waiver of preemption for other states that have adopted California's standards, if EPA grants California a waiver.)

Faced with severe air pollution problems, especially in Los Angeles and the San Joaquin Valley, California has regularly developed more stringent standards for motor vehicle emissions than those required by federal law. In order to impose its standards, the state has requested and been granted Section 209(b) waivers at least 54 times since 1967.[4] Using Section 209(b) waivers, California has served as a laboratory for the demonstration of cutting-edge emission control technologies, which, after being successfully demonstrated there, were adopted in similar form at the national level. Catalytic converters, cleaner fuels, and numerous other advances were introduced in this way. Currently, waivers allow California to require that a portion of each manufacturer's sales meet Zero Emission Vehicle (ZEV) and Partial ZEV requirements, which has stimulated the sale of electric and hybrid vehicles.

CALIFORNIA'S GREENHOUSE GAS REQUIREMENTS

On July 22, 2002, California became the first state to enact legislation requiring reductions of greenhouse gas (GHG) emissions from motor vehicles. The legislation, AB 1493, required the California Air Resources Board (CARB) to adopt regulations requiring the "maximum feasible and cost-effective reduction" of GHG emissions from any vehicle whose primary use is noncommercial personal transportation. GHGs are defined by the state as carbon dioxide, methane, nitrous oxide, hydrofluorocarbons, perfluorocarbons, and sulfur hexafluoride, but for the purpose of this regulatory program, only the first four of these are subject to control. The reductions were to apply to motor vehicles manufactured in the 2009 model year and thereafter.

Under this authority, CARB adopted regulations September 24, 2004, requiring gradual reductions in fleet average GHG emissions until they are about 30% below the emissions of the 2002 fleet in 2016.[5] As illustrated in **Figure 1**, the regulations set separate standards for two classes of vehicles. The first class consists of all passenger cars, plus light duty trucks and SUVs weighing 3,750 lbs. or less; these vehicles must reduce emissions by an average of 36.5% between 2009 and 2016. The second group consists of light trucks and passenger vehicles over 3,750 lbs., which must reduce emissions 24.4% over the same time period.

The regulations require reductions in fleet averages, rather than compliance by individual vehicles. They provide substantial flexibility, including credit generation from alternative fuel vehicles and averaging, banking, and trading of credits within and among manufacturers. Credits—and debits for any year in which a manufacturer exceeds the standards—must be equalized within five years of their generation, with the first equalization required in 2014. Thus, manufacturers would not be subject to penalties for failure to meet the standards until 2014 at the earliest.[6] Following adoption of these regulations by CARB, they were subjected to public comment and legislative review, and CARB submitted a request to U.S. EPA, December 21, 2005, for a preemption waiver under Section 209(b).

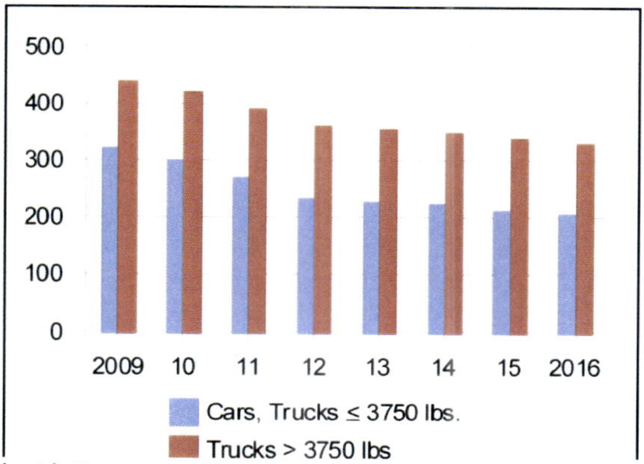

Source: California Air Resources Board.

Figure 1. California GHG Emission Requirements (grams/mile, CO2 equivalent)

EPA'S RESPONSE TO THE WAIVER REQUEST AND RESULTING LITIGATION

EPA took two years, from late 2005 to late 2007, to respond to California's waiver request for its motor vehicle GHG emission standards. The agency's long response time was doubtless the result of intense debate over the waiver request within the Bush Administration, which from its arrival in Washington had opposed regulatory approaches to reducing GHG emissions. In addition, the agency was waiting for the U.S. Supreme Court to decide a case it had accepted that squarely presented the issue of whether GHGs are "air pollutants" under the CAA, and thus subject to EPA's regulatory authority. This case began when EPA, in 2003, denied a petition asking it to regulate GHG emissions from new motor vehicles under CAA section 202(a).[7] The agency concluded it lacked authority under the CAA to regulate motor vehicle emissions based on their climate change effects. In its April 2, 2007 decision in *Massachusetts v. EPA*,[8] the Supreme Court resolved this authority issue, finding 5-4 that—

> The Clean Air Act's sweeping definition of "air pollutant" includes "*any* air pollution agent or combination of such agents, including *any* physical, chemical ... substance or matter which is emitted into or otherwise enters the

> ambient air.... " ... Carbon dioxide, methane, nitrous oxide, and hydrofluorocarbons are without a doubt "physical [and] chemical ... substance[s] which [are] emitted into ... the ambient air." The statute is unambiguous.[9]

Thus, the Court's majority had no doubt that the CAA gives EPA authority to regulate GHGs from new motor vehicles, although the specifics of such regulation are subject to agency discretion. (See CRS Report RS22665, *The Supreme Court's Climate Change Decision: Massachusetts v. EPA*, by Robert Meltz.)

Following this decision, EPA announced that it would consider the California waiver request. The agency held public hearings on May 22, 2007, in Arlington, VA, and on May 30, 2007, in Sacramento, CA. Under pressure from California's Senator Boxer, who chairs the Environment and Public Works Committee,[10] and other California leaders, including Governor Schwarzenegger and Attorney General Brown,[11] EPA Administrator Johnson announced that he would decide whether to grant the waiver request by the end of 2007.[12]

During the public comment period, the agency received more than 60,000 comments, the vast majority of them urging it to grant the waiver. Support came from environmental groups, the Manufacturers of Emission Controls Association, the National Association of Clean Air Agencies (which represents state and local air pollution control departments), and a number of governors. As will be discussed further below, 14 other states and the District of Columbia have adopted regulations identical to California's, and 2 others have announced their intention to do so, but their ability to implement the regulations depends on California first being granted a waiver.[13] Thus, many of them have weighed in in support of the waiver request.

The auto industry and the U.S. Department of Transportation (DOT), among others, opposed a waiver grant. The auto industry maintains that there is effectively no difference between California and federal emission standards in their impact on criteria air pollutants (ozone, in particular), that the benefits of the GHG regulations are "zero," and that emissions from California's auto fleet will actually increase as a result of the regulations as consumers keep older, higher-emitting cars longer.[14]

On December 19, 2007, EPA announced its decision. EPA Administrator Stephen Johnson wrote California Governor Arnold Schwarzenegger to say, "I have decided that EPA will be denying the waiver and have instructed my staff to draft appropriate documents setting forth the rationale for this denial in further detail.... " According to press reports, the decision was taken against the unanimous advice of the agency's technical and legal staffs.[15] His staff did subsequently draft a decision document, which the Administrator signed on February 29, 2008.[16] It was published in the March 6, 2008 *Federal Register*,[17] and will be referred to herein as the March 6 decision.

Table 1. States Adopting California's Mobile Source GHG Standards

State	2006 Population	Legislation/Regulation
Arizona	6,166,318	Executive Order 2006-13, September 8, 2006
California	36,457,549	AB 1493, July 22, 2002
Connecticut	3,504,809	Public Act 04-84, May 4, 2004
District of Columbia	581,530	Law 17-0151, March 19, 2008
Florida	18,089,888	Executive Order 07-127, July 13, 2007
Maine	1,321,574	Amendments to Chapter 127, December 19, 2005
Maryland	5,615,727	Senate Bill 103, April 24, 2007
Massachusetts	6,437,193	Amendments to the state's LEV regulations, December 30, 2005
New Jersey	8,724,560	P.L. 2003, Chapter 266, January 14, 2004
New Mexico	1,954,599	Executive Order 2006-69, December 28, 2006
New York	19,306,183	Chapter III, Subpart 218-8, November 9, 2005
Oregon	3,700,758	Regulations (Division 257; OAR 340-256-0220; and Division 12), June 22, 2006
Pennsylvania	12,440,621	Amendments to Title 25, Chapters 121 and 126, December 9, 2006
Rhode Island	1,067,610	Air Pollution Control Regulation No. 37, December 22, 2005
Vermont	623,908	Amendments to Subchapter XI, November 7, 2005
Washington	6,395,798	House Bill 1397, May 6, 2005
Total	132,388,625	

Source: Pew Center on Global Climate Change for information and links to state regulations, at http://www.pewclimate.org/what_s_being_done/in_the_states/vehicle_ghg_standard.cfm, U.S. Census Bureau for population data. As of February 28, 2008, the Pew Center also listed Colorado and Utah as having announced their intention to adopt California's standards, although neither state had formally adopted legislation or regulations as of that date.

It came as no surprise that the Administrator's denial of California's waiver petition was challenged in court. The first petitions for review were filed soon after the December 19 letter by California, 15 other states, and environmental groups, arguing that the letter itself constituted final agency action and was thus ripe for review. With the issuance of the March 6 decision document, these suits based on the EPA letter were dismissed (by the court or by stipulation) and replaced, on May 5, 2008, by a suit in the D.C. Circuit challenging that document.[18] Petitioners in *California v. EPA* are California, 18 other states, and numerous environmental groups. Most of the California congressional delegation, including Speaker of the House Nancy Pelosi and Senators Boxer and Feinstein, are participating in the case as amici on behalf of the petitioners. The legal arguments of petitioners and respondent EPA are summarized in the "Waiver Criteria" section of this report, under the subsection discussing the "Compelling and Extraordinary Conditions" waiver criterion that is the focus of these arguments. Briefing in the case is now underway; a decision is likely in spring, 2009 (unless mooted by the Obama Administration EPA's possible reversal of the denial).

ACTIONS BY OTHER STATES

As noted above, California is the only state permitted to adopt more stringent emission standards under the waiver provision of Section 209(b). CAA Section 177, however, provides that any state with an EPA-approved State Implementation Plan—every state except Hawaii, North Dakota, and South Dakota—"may adopt and enforce for any model year standards relating to control of emissions from new motor vehicles or new motor vehicle engines" provided: (1) that the standards are identical to standards for which California has been granted a waiver; and (2) that California and such state have adopted the standards at least two years before the commencement of the model year to which the standards apply. Relying on this authority, and presuming that California will be granted a waiver, 14 other states (Arizona, Connecticut, Florida, Maine, Maryland, Massachusetts, New Jersey, New Mexico, New York, Oregon, Pennsylvania, Rhode Island, Vermont, and Washington) and the District of Columbia have adopted or announced their intention[19] to adopt California's greenhouse gas emission controls. Including California, these states account for 44% of the total U.S. population

(**Table 1**).[20] Thus, the stakes involved (both the environmental consequences and the potential impact on the auto industry) go well beyond California.

WAIVER CRITERIA

As noted earlier, Section 209(b) says that the EPA Administrator "shall ... waive" the prohibition on state emission standards "if the State determines that the State standards will be, in the aggregate, at least as protective of public health and welfare as applicable Federal standards." The section adds:

> No such waiver shall be granted if the Administrator finds that-
>
> (A) the determination of the State is arbitrary and capricious,
>
> (B) such State does not need such State standards to meet compelling and extraordinary conditions, or
>
> (C) such State standards and accompanying enforcement procedures are not consistent with section 202(a) of this part.

As originally enacted in 1967, Section 209(b) was worded differently. The waiver process did not involve, as it does today, an initial determination of protectiveness by California that EPA can reject only on an "arbitrary and capricious" standard that is deferential to the state. This bow to California was added by the 1977 amendments, the legislative history of which stresses that the changes made in Section 209(b) were "intended to ratify and strengthen the California waiver provision and to affirm the underlying intent of that provision, i.e., to afford California the broadest discretion possible in selecting the best means to protect the health of its citizens and the public welfare."[21]

California's Justification of Its Waiver Request

In requesting a waiver, California argued that its standards met each of the CAA criteria: that they were at least as protective as applicable federal standards, because there are no federal GHG standards; that air pollution generally, and climate change in particular, present numerous conditions to the state that are compelling and extraordinary; and that the standards and enforcement procedures are consistent with Section 202(a), based on previous EPA interpretation of the consistency requirement.[22]

Of particular importance (see following section) is the state's case that it faces compelling and extraordinary conditions. These include the potential of rising sea levels that would bring increased salt water intrusion to its limited supplies of water; diminishing snow pack that would also threaten its limited water supply; and higher temperatures that would exacerbate the state's ozone nonattainment problem, which is already the worst in the nation.[23] But the state argues that it need not demonstrate that it faces unique threats from GHG emissions, since EPA must consider not just the GHG controls, but the state's entire motor vehicle emissions program in evaluating the waiver request. According to the state: "The relevant inquiry under section 209(b)(1)(B) is whether California needs its own emission control program to meet compelling and extraordinary conditions, not whether any given standard is necessary to meet such conditions."[24]

The Decision Document's Argument for Waiver Denial

In its March 6 decision document denying California's waiver request, EPA explained that its denial was based solely on CAA Section 209(b)(1)(B), the "compelling and extraordinary conditions" criterion. It did not address the other two criteria above.

EPA concluded that California did *not* need its GHG emission standards to meet compelling and extraordinary conditions. At the outset, this argument required EPA to reverse its historical position that the "State standards" mentioned in Section 209(b)(1)(B) embrace California's *entire* motor vehicle emissions control program, not merely the standards that are the subject of the waiver request. This longstanding agency position was based partly on statutory text: the phrase "such State standards" in 209(b)(1)(B) seems to refer back to "State standards ... in the aggregate." Lending further support is 1977 legislative history, which speaks to a situation where EPA might want to approach an air pollution problem by cranking down on the allowable emissions for one pollutant, but California might prefer to address the problem by cutting down on another. "To deal with such a situation," says the 1977 House report, "the Committee amendment requires the Administrator of EPA to grant a waiver for the *entire* set of California standards."[25]

For California's GHG regulations, however, EPA found the opposite – that it was "appropriate" to consider the need to meet compelling and extraordinary conditions with reference to the GHG regulations *alone.* The reason for the switch, EPA said, was that GHG standards address a *global* problem, whereas in the past California's motor vehicle program, and its waiver requests, had only addressed *local and regional* problems. Local and regional air pollution problems are rooted in local causes, such as thermal inversions and the large number of vehicles in California. In contrast, GHG concentrations are essentially uniform across the globe, and are not uniquely connected with causal factors in California. Based on this lack of nexus between local conditions and global GHG concentrations, California does not, EPA concluded, "need" its GHG emission standards to meet the compelling and extraordinary condition of climate change.

Alternatively, EPA argued, the impacts of climate change in California are simply not the kind of impacts that Congress had in mind by the phrase "compelling and extraordinary conditions." The 1967 legislative history of Section 209(b), the agency points out, refers to California's "peculiar local conditions" and "unique problems," suggesting to EPA that to be compelling and extraordinary, the impacts of climate change in California must be significantly different from those in other states. To prove such uniqueness, California had noted in its petition that it faced greater temperature increases from climate change than the rest of the nation, that it has the largest agriculture-based economy of any state, the largest coastal population, and so on. EPA was not persuaded, concluding that global climate change will affect the nation and the world in ways very similar to California.

Positions of the Parties in the Waiver Denial Litigation

As noted, the active suit against the waiver denial is *California v. EPA*, filed May 5, 2008, by California, other states, and environmental groups. In their joint brief, these petitioners argue first that EPA erred in not applying Section 209(b)(1)(B) to California's motor vehicle emissions program as a whole. Petitioners reassert the textual argument – that "such State standards" in Section 209(b)(1)(B) refers back to the phrase "State standards ... in the aggregate." They also stress the longevity of EPA's historical position that the entire California motor vehicle emissions program must be assessed.[26]

Petitioners then make a typical *Chevron* step one/*Chevron* step two argument. By way of background, the Supreme Court in *Chevron U.S.A. Inc. v. NRDC* set out what has become the canonical framework for judicial review of federal agency action.[27] Under *Chevron*, the reviewing court first asks whether "Congress has directly spoken to the precise question at issue."[28] If so, the court is to "give effect to the unambiguously expressed intent of

Congress."[29] On the other hand, if the statute is found to be "silent or ambiguous" on the question, the court must proceed to Chevron step two and ask whether the agency's interpretation was "permissible."[30] A petitioner challenging agency action will, if possible, argue that the agency interpretation fails under step one, since under step two the court generally must accord the agency interpretation a considerable degree of deference.

Petitioners' "*Chevron* step one" argument is that EPA's exclusion of climate change impacts in California from "compelling and extraordinary conditions" because they are not local/regional and not unique to the state violates the plain meaning of Section 209(b)(1)(B). Petitioners argue that the fact that smog reduction was the dominant concern in 1967, when the section was enacted, does not mean that the broad language of Section 209(b) should be confined to that situation today. To buttress this view, petitioners argue that construing "compelling and extraordinary" to require that conditions in California be unique is contradicted by CAA Section 177, discussed above. How, they ask, can it make sense to allow other states to adopt California's standards if California's authority is limited to addressing problems unique to that state?

Petitioner's "*Chevron* step two" argument is that even assuming EPA's interpretation of compelling and extraordinary conditions survives *Chevron* step one, it is not a "permissible" interpretation. It is not permissible, they contend, because it conflicts with the Supreme Court's view in *Massachusetts v. EPA* that climate change is within the CAA's scope notwithstanding that Congress did not have it in mind when the CAA was enacted. Moreover, the deference normally accorded an agency's interpretation under *Chevron* step two should be at a minimum here, petitioners argue, because EPA's waiver denial infringes on state sovereignty and reverses its longstanding practice of assessing California's vehicle emissions control program in the aggregate.

In response to petitioners' arguments, the Bush Administration EPA argued, in one of its last legal briefs filed January 9, 2009, that its waiver rejection should be upheld. Its positions and supporting arguments in the brief track those in its decision document – and are all positioned under the usually deferential *Chevron* step two. As for the threshold issue, EPA argues that its decision to review California's GHG regulations separately was a reasonable interpretation of an ambiguous statute. EPA then argues that it reasonably interpreted the reference to compelling and extraordinary conditions in Section 209(b)(1)(B) to focus on local conditions, and on whether those conditions are sufficiently different from the rest of the country. Under that interpretation, it contended, California had not demonstrated that its standards were "needed" to meet such conditions.

The fate of the *California v. EPA* litigation hinges on what action, if any, the newly arrived Obama Administration officials take at the EPA. EPA could reverse the Bush Administration EPA's waiver denial, making the present suit moot but inviting a judicial challenge to the reversal in a new suit by the auto manufacturers. Alternatively, the Obama Administration EPA could wait until the D.C. Circuit rules (putting it in the probably unacceptable position of defending the waiver denial at oral argument), or until Congress mandates that the waiver be granted (mooting any litigation). The complexities of each of these courses of action are explored in the final section of this report.

PREVIOUS WAIVER REQUESTS

As noted earlier, California has requested waivers under Section 209(b) on many occasions. A precise count of the number of such requests is difficult to determine, according to EPA's Office of Transportation and Air Quality (OTAQ), in large part because the nature of such requests varies. The state has requested waivers for new or amended standards on at least 55 occasions; on at least 42 other occasions, the state has requested "within the scope" determinations (i.e., a request that EPA rule on whether a new regulation is within the scope of a waiver that the agency has already issued). Adding all of these together, one might say that there have been nearly 100 waiver requests.[31]

Of these, all were granted in whole or in part. EPA has repeatedly found, as recently as September 2008, that California faces compelling and extraordinary conditions (as to conventional pollutants, not climate change) and needs its own standards to meet these conditions.[32] In general, as the Administrator stated in a 1975 waiver determination:

> These provisions must be read in the light of their unusually detailed and explicit legislative history.... Congress meant to ensure by the language it adopted that the Federal government would not second-guess the wisdom of state policy here.... Sponsors of the language eventually adopted referred repeatedly to their intent to make sure that no "Federal bureaucrat" would be able to tell the people of California what auto emission standards were good for them, as long as they were stricter than Federal standards.[33]

In arguing thus, the Administrator foreshadowed the House Interstate and Foreign Commerce committee report on the 1977 CAA Amendments, which revisited and strengthened California's position in seeking a waiver. The report, accompanying amendments to Section 209(b) that gave the subsection its current form, states:

> The Administrator, thus, is not to overturn California's judgment lightly. Nor is he to substitute his judgment for that of the State. There must be clear and compelling evidence that the State acted unreasonably in evaluating the relative risks of various pollutants in light of the air quality, topography, photochemistry, and climate in that State, before EPA may deny a waiver.[34]

Although EPA had never denied a request outright before the GHG waiver, on at least six occasions prior to the 1977 CAA amendments, the agency granted a waiver in part, while denying or delaying the effective date of other parts of the request on feasibility grounds.[35] In 1975, the Administrator denied a waiver for the 1977 model year, but granted it for 1978.[36] Since the 1977 amendments, there was at least one instance in which EPA made a determination that California's requirements were feasible in part, granting a waiver for the 2007 through 2011 model years, but making no decision for model years after that.[37]

Related Litigation

Aside from litigation over EPA's denial of California's request for a CAA preemption waiver, there is active litigation over state regulation of mobile source GHG emissions raising non-CAA preemption and other legal theories. This litigation, filed by auto dealers, trade associations, and auto manufacturers, seeks to prevent California and other states from implementing the California mobile source GHG standards *even if the EPA waiver denial is overturned by the courts or reversed by EPA.* Suits are pending in four federal judicial circuits—not coincidentally, the circuits containing most of the states that have adopted the California GHG controls. Courts addressing this litigation have not doubted that without a California waiver, state regulation of GHG emissions from motor vehicles is preempted by the CAA, and the non-CAA litigation is moot.

The chief non-CAA preemption theory in this litigation is based on the Energy Policy and Conservation Act (EPCA), the authority under which the

National Highway Traffic Safety Administration (NHTSA) establishes corporate average fuel economy standards ("CAFE standards").[38] Enacted in 1972, EPCA today requires NHTSA to prescribe separate fuel economy standards for passenger and non-passenger automobiles beginning with model year 2011, to achieve a combined fuel economy average for model year 2020 of at least 35 miles per gallon.[39] More pertinent here, EPCA preempts states from adopting laws "related to" the federal fuel economy standards.[40] The auto industry argues that this preemption is fatal to state regulation of GHG emissions from cars. The argument runs that the only known way to reduce GHG emissions is to improve gas mileage, so that a state regulation of auto GHG emissions is a law "related to" the federal emission standard, hence invalid.

Two decisions on the merits have been handed down so far, both rejecting the non-CAA preemption theories, including under EPCA. In the first, *Green Mountain Chrysler Plymouth Dodge Jeep v. Crombie*,[41] the district court ruled that the relationship between Vermont's California-identical GHG standards and EPCA was better analyzed as an interplay between two federal statutes (EPCA and the CAA), rather than as a federal-state preemption question. So viewing the matter, the court pointed out that NHTSA has consistently treated EPA-approved California emissions standards as constituting "other motor vehicle standards of the Government," which EPCA says NHTSA must consider when setting CAFE standards.[42] This suggests that EPCA was meant to coexist with the CAA, rather than supersede it. Moreover, noted the court, in a related context the *Massachusetts v. EPA* decision saw the CAA and EPCA CAFE provisions as harmonious.[43] Thus, the court found the CAA section 209/EPCA relationship to be one of overlap, not conflict. Despite its conclusion that preemption doctrine did not apply, the court also did a preemption analysis, finding that Vermont's GHG standards were preempted neither by EPCA nor as an intrusion upon the foreign policy authority of the United States. An appeal is pending.

In the second decision, *Central Valley Chrysler Jeep, Inc. v. Goldstene*,[44] a district court similarly rejected claims that California's regulation of GHG emissions from cars and trucks was precluded by EPCA, preempted by EPCA, and preempted as an intrusion on federal authority over foreign policy.[45] An appeal is pending in this case as well.

The legal theories pressed in the *Crombie* and *Goldstene* cases are similar to those in two Rhode Island suits, consolidated as *Lincoln Dodge, Inc. v. Sullivan,*[46] challenging that state's adoption of the California standards. Recently, the district court held that the claims of the auto manufacturers and trade associations in this case were barred by collateral estoppel, a legal doctrine that prohibits parties from relitigating issues they have already adjudicated (as these plaintiffs had done in *Crombie* and *Goldstene*).[47] The Rhode Island auto dealers, by contrast, had themselves never raised the issues in the case, and thus were held to be viable plaintiffs, allowing the case to proceed. In a fourth preemption case, New Mexico's adoption of the California GHG standards has been challenged as preempted under EPCA in *Zangara Dodge, Inc. v. Curry*.[48]

COMMENTS AND CONCLUSIONS

In considering whether to reverse the Bush Administration's stance on the California waiver, the incoming Administration faces several important decisions regarding the potential for greenhouse gas regulations. EPA has now received at least eight petitions asking the agency to regulate GHGs from different sources, including cars and trucks, aircraft, ocean-going ships, nonroad engines and equipment (such as locomotives, construction equipment, farm equipment, forklifts, harbor craft, and lawn and garden equipment), and fuels—in addition to the California waiver request. The agency also faces lawsuits seeking to force it to regulate GHGs from stationary sources, including power plants, petroleum refineries, and the Portland cement industry. A decision on any of the petitions is likely to be seen as a precedent for the others, and would have implications for the agency's position in the pending waiver-denial litigation.

The incoming Obama Administration is sympathetic to greenhouse gas controls. At the same time, both Congressional leaders and President Obama have stated their intention to make the passage of economy-wide GHG controls a priority. A decision on the California waiver and the other pending petitions needs to be viewed, therefore, in the context of its potential impact on such legislation as well as the precedent it might set for GHG regulations under existing CAA authority.

From statements made during the election campaign, President Obama appears poised to reverse EPA's denial of the waiver request,[49] and since his inauguration, the President has ordered EPA to review the decision. In general, a regulation (or "rule") can be reversed by following the same procedural steps that were taken to promulgate it. The waiver decision is not a regulation—it was never formally proposed in the *Federal Register*, and CAA Section 307 does not treat decisions under Section 209(b) as being subject to the administrative procedures that it sets out for regulatory decisions – but the same general principal (i.e., following the same procedural steps the agency took to make the original decision in order to reverse it) presumably applies. A waiver decision, states Section 209(b), requires only "notice and opportunity for a public hearing" before it is made. In reaching the March 6, 2008 decision, the EPA Administrator requested public comment on whether he should grant the waiver, held two public hearings, and six months later announced that he had decided not to grant a waiver, stating his reasons.

Thus, reversing the decision could be relatively simple – less complicated than reversing a formal regulation. It would not require a formal proposal, since the original decision was not proposed. Instead, the Administrator can take action following notice and opportunity for a public hearing. The final step would be to set forth a rationale for the decision's reversal. This could be relatively straightforward, as well: the justification for granting a waiver was prepared by EPA staff, before Administrator Johnson discarded it and changed course in December 2007.[50] Thus, a reversal by the new EPA Administrator might be the quickest way to address the issue.

California and EPA are already pursuing this tack: a January 21, 2009 letter to EPA from the California Air Resources Board formally requests EPA to reconsider its denial, and based on the Board's arguments (similar to those pressed by California in the litigation), grant the waiver.[51] In her confirmation hearing, Administrator Jackson stated that she would reconsider the waiver request, and President Obama requested the agency to do so January 26. On February 6, the agency announced that it would hold a hearing on California's request March 5, 2009, in Arlington, VA, and that it would take public comment through April 6.

Given the determination to proceed administratively, there may be little need for Congress to involve itself in the decision; nevertheless, interest in the issue is expected to remain high until it is resolved.

End Notes

[1] CAA § 209(a), 42 U.S.C. § 7543(a). See also S.Rept. 91-1196 (1970), p. 32.

[2] 42 U.S.C. § 7543(b).

[3] As will be discussed in greater detail below, there are three conditions placed on the grant of such waivers: The Administrator is to deny a waiver if he finds: (1) that the state's determination is arbitrary and capricious; (2) that the state does not need separate standards to meet compelling and extraordinary conditions; or (3) that the state's standards and accompanying enforcement procedures are not consistent with Section 202(a) of the Act.

[4] Personal communication, U.S. EPA Office of Transportation and Air Quality, July 20, 2007.

[5] A table showing the mandated reductions year-by-year can be found in CARB's *Regulations to Control Greenhouse Gas Emissions from Motor Vehicles, Final Statement of Reasons*, August 4, 2005, p. 8 at http://www.arb.ca.gov/regact/grnhsgas/fsor.pdf.

[6] California Air Resources Board, Regulations to Control Greenhouse Gas Emissions from Motor Vehicles; Request for Waiver of Preemption Under Clean Air Act Section 209(b), December 21, 2005, Attachment 2, Support Document, p.2, available at http://www.arb.ca.gov/cc/docs/att2_support.pdf. Hereafter referred to as "Support Document."

[7] 42 U.S.C. § 7521(a).

[8] 549 U.S. 497 (2007).

[9] Id. at 528-529 (emphasis in original).

[10] At a May 22, 2007 hearing, for example, Senator Boxer stated, "EPA already has all the authority it needs to begin regulating greenhouse gas emissions from motor vehicles now. The Supreme Court's landmark decision has now cleared the way. The time to act is now. The clearest example of this point is the case for the California waiver. ... Further delay in this matter is simply unacceptable." See Opening Statement of Senator Barbara Boxer, U.S. Senate, Committee on Environment and Public Works, Hearing on "Examining the Case for the California Waiver," May 22, 2007, at http://epw.senate.gov/public/index.cfm?FuseAction=Hearings.Statement&Statement_ID=39 508511-fd9e- 469b-80af-faaf843f6696.

[11] See "California Attorney General to File Lawsuit if EPA Fails to Act on Waiver Past October 25," *Daily Environment Report*, May 23, 2007, p. A-13.

[12] Testimony of Stephen L. Johnson, Administrator, U.S. EPA, before the Senate Committee on Environment and Public Works, July 26, 2007, at http://epw.senate.gov/public/index.cfm?FuseAction=Files.View&FileStore_id=1a49cc26-6d6b-4f55-9eb4-759b7e0e039c.

[13] The 14 states are Arizona, Connecticut, Florida, Maine, Maryland, Massachusetts, New Jersey, New Mexico, New York, Oregon, Pennsylvania, Rhode Island, Vermont, and Washington. Under Section 177 of the Act, states that have nonattainment or "maintenance" areas can adopt California's emission standards for mobile sources in lieu of federal standards. Every state except Hawaii, North Dakota, and South Dakota would be eligible to adopt California's standards under this so-called "piggyback" provision. Thus, there is broad interest in the California waiver decision and more at stake than would be the case if only California had adopted the regulations.

[14] Alliance of Automobile Manufacturers, "California Waiver Request," presentation materials from U.S. EPA public hearing, Sacramento, CA, May 30, 2007.

[15] "EPA Chief Denies Calif. Limit on Auto Emissions," *Washington Post*, December 20, 2007, p. A1. Documents shown to, and transcribed by, congressional staff have included numerous statements by senior EPA staff recommending that the Administrator grant the waiver; and the Administrator did not identify any staff recommendation suggesting denial. See U.S. Senate, Committee on Environment and Public Works, Hearings, January 24, 2008, and February 27, 2008. Additional detail confirming the above was provided as the result of an investigation by the House Oversight and Government Reform Committee. See "White House Involved in California Waiver Denial," May 19, 2008, at [http://oversight.house.gov/investigations.asp?start=25&id=121].

[16] Related materials can be found at http://www.epa.gov/otaq/ca-waiver.htm.

[17] 73 *Federal Register* 12013.

[18] State of California v. U.S. EPA, No. 08-1178 (D.C. Cir. filed May 5, 2008).

[19] In some cases, only one branch of government (e.g., the Governor, through Executive Order) has ordered the adoption of the California GHG standards. Without reviewing each state's regulatory process, it is unclear to CRS whether, in such cases, the state can be considered to have adopted the standards.

[20] Colorado and Utah can perhaps be added to this list: in Utah's case, the state has joined a regional group that has pledged to adopt the California standards; in Colorado, the Governor has released a Climate Action Plan that includes the California standards. Neither state has issued regulations as of this writing, however.

[21] U.S. Congress, House Interstate and Foreign Commerce Committee, *Clean Air Act Amendments of 1977*, May 12, 1977, H.Rept. 95-294, pp. 301-302.

[22] The state concluded that, since there are no federal test procedures that measure GHGs for climate change purposes, inconsistency with Section 202(a) can only be shown if there is inadequate lead time to permit the development of technology to meet the regulatory requirements. The state based the standards on technologies already available on vehicles or demonstrated by auto companies and component suppliers, so it concluded that the standards were consistent with the need for adequate lead time. See CARB, Support Document accompanying the December 21, 2005 Waiver Request, previously cited, pp. 19-43.

[23] CARB, Support Document, p. 18.

[24] Support Document, p. 15.

[25] H.Rept. 95-294, p. 302 (emphasis added).

[26] *See, e.g.,* 49 *Federal Register* 18,887, 18,889-90 (May 3, 1984).

[27] 467 U.S. 837 (1984).

[28] Id. at 842.

[29] Id. at 843.

[30] Id.

[31] Personal communication, U.S. EPA, Office of Transportation and Air Quality, July 20, 2007. California has also submitted about 10 waiver requests for non-road vehicles and engines under Section 209(e). These form a third category.

[32] 73 *Federal Register* 52042, September 8, 2008.

[33] 40 *Federal Register* 23103, May 28, 1975.

[34] U.S. House of Representatives, Committee on Interstate and Foreign Commerce, Clean Air Act Amendments of 1977, H.Rept. 95-294, May 12, 1977, pp. 301-302.

[35] According to EPA, the dates were May 6, 1969 (34 *Federal Register* 7348), April 30, 1971 (36 *Federal Register* 8172), April 25, 1972 (37 *Federal Register* 8128), April 26, 1973 (38 *Federal Register* 10317), November 1, 1973 (38 *Federal Register* 30136), and July 18, 1975 (40 *Federal Register* 30311).

[36] 40 *Federal Register* 30311, July 18, 1975.

[37] 71 *Federal Register* 78190, December 28, 2006.

[38] The fuel economy provisions of EPCA are codified at 49 U.S.C. §§ 32901-32919.

[39] 49 U.S.C. § 32902(b)(2)(A).

[40] 49 U.S.C. § 32919.

[41] 508 F. Supp. 2d 295 (D. Vt. 2007).

[42] 49 U.S.C. § 32902(f).

[43] 549 U.S. at 532.

[44] 529 F. Supp. 2d 1151 (E.D. Cal. 2007).

[45] In 2006, the district court dismissed claims under the Dormant Commerce Clause and Sherman Antitrust Act.

[46] No. 1:06-CV-00070 (D.R.I. filed February 13, 2006).

[47] November 24, 2008.

[48] No. 1:07-CV-01305 (D.N.M. filed December 27, 2007).

[49] "Obama Likely to Take Action on Regulating Emissions, Clearing California Waiver Request," *Daily Environment Report*, November 13, 2008, p. A-1. Of course, the decision is made by the EPA Administrator, not by the President. Since the Bush Administration was criticized for having – at least to outward appearances – pressured the EPA Administrator to deny the waiver, President Obama and his Office of Management and Budget may be hesitant to become too overtly involved in EPA's reconsideration of its waiver denial.

[50] The history of the decision is known in some detail as the result of an investigation by the House Oversight and Government Reform Committee. See "White House Involved in California Waiver Denial," May 19, 2008, at http://oversight.house.gov/investigations.asp?start=25&id=121.

[51] Letter from Mary Nichols, Director, California Air Resources Board, to EPA Administrator-Designate Lisa P. Jackson, dated January 21, 2009, available at http://www.arb.ca.gov/newsrel/arbwaiverrequest.pdf .

In: Climate Change Litigation and Law
Editors: Jean-François Masson

ISBN: 978-1-60876-089-3

Chapter 3

CLIMATE CHANGE: FEDERAL LAWS AND POLICIES RELATED TO GREENHOUSE GAS REDUCTIONS

Brent D. Yacobucci and Larry Parker
Energy and Environmental Policy

SUMMARY

Climate change is viewed as a global issue, but proposed responses generally require action at the national level. In 1992, the United States ratified the United Nations Framework Convention on Climate Change (UNFCCC), which called on industrialized countries to take the lead in reducing greenhouse gases. Over the past 16 years, a variety of voluntary and regulatory actions have been proposed or undertaken in the United States, including monitoring of electric utility carbon dioxide emissions, improved appliance efficiency, and incentives for developing renewable energy sources. This report provides background on the evolution of U.S. climate change policy, from ratification of the UNFCCC to the George W. Bush Administration's 2001 rejection of the Kyoto Protocol to the present. Recent federal court decisions—most notably the Supreme Court's 2007 decision in *Massachusetts v. EPA* that the Environmental Protection Agency has the authority to regulate motor vehicle greenhouse gas emissions under the Clean Air Act—have raised the issue of whether EPA should directly regulate greenhouse gases. This report focuses on major regulatory programs that monitor or reduce greenhouse gas emissions, along with their estimated effect on emissions levels.

The George H. W. Bush, Clinton, and George W. Bush Administrations largely relied on voluntary initiatives to reduce the growth of greenhouse gas emissions. This focus was particularly evident in the current Administration's 2006 Climate Action Report (CAR), submitted under the provisions of the UNFCCC. Of roughly 50 programs summarized in the 2006 CAR, seven were described as "regulatory." However, this small subset of the total U.S. effort accounts for a large share of greenhouse gas emission reductions achieved over the past decade-and-a-half. In general, these efforts were established and implemented in response to concerns other than climate change, such as energy efficiency and air quality.

The Energy Policy Act of 2005 (P.L. 109-58) included provisions indirectly related to greenhouse gas emissions, such as energy efficiency and renewable energy. The Energy Independence and Security Act of 2007 (P.L. 110-140) addresses renewable energy and conservation, but also includes provisions specifically on climate change. These include a requirement for the use of renewable fuels with lower lifecycle greenhouse gas emissions than petroleum fuels, and the establishment of an Office of Climate Change and Environment in the Department of Transportation to implement research on mitigating the causes and addressing the effects of climate change on transportation. In June 2008, the Senate considered a bill (S. 3036) to establish an economy-wide cap-and-trade system to reduce greenhouse gas emissions. However, after discussion, a cloture motion on this bill failed, and the bill was tabled.

While some provisions in energy laws enacted over the past 16 years have led to lower greenhouse gas emissions or addressed climate change directly, other provisions in those same laws have almost certainly resulted in higher emissions. To date, no energy law has had reducing greenhouse gas emissions as the main organizing principle. Energy-related activities are responsible for about 86% of the country's greenhouse gas emissions, and 98% of its carbon dioxide emissions. Climate change policy directed at reducing greenhouse gas emissions must address energy supply and consumption and, thus, be integrated with energy policy. This will be a pivotal challenge to the 111th Congress's and the incoming Administration's anticipated efforts to enact legislation to limit greenhouse gas emissions.

INTRODUCTION

Climate change is viewed as a global issue, but proposed responses generally require action at the national level. In 1992, the United States ratified the United Nations Framework Convention on Climate Change (UNFCCC) which called on industrialized countries to take the lead in reducing greenhouse gases. Over the past 16 years, a variety of voluntary and regulatory actions have been proposed or undertaken in the United States, including monitoring of utility carbon dioxide emissions, improved appliance efficiency, and incentives for developing renewable energy sources.

In 2001, President George W. Bush rejected the Kyoto Protocol to the UNFCCC, which called for legally binding commitments by developed countries to reduce their greenhouse gas emissions.[1] He also rejected the concept of mandatory emissions reductions. Since then, the Bush Administration has focused U.S. climate change policy on voluntary initiatives to reduce the growth in greenhouse gas emissions. This focus was particularly evident in the Administration's 2006 Climate Action Report (CAR) submitted under the provisions of the UNFCCC. Of the roughly 50 programs summarized in the 2006 CAR, only seven were described as "regulatory."[2] These regulatory programs were generally implemented to achieve energy or environmental goals other than the reduction of greenhouse gas emissions, but produced a concomitant emissions reduction. In this sense, they could be considered the results of a "no regrets"[3] policy in which climate change effects resulting from related air quality and energy policies are included in the decision-making process on new or modified rules.

However, indirect regulation and "no regrets" policies may be supplanted by direct regulation of greenhouse gas emissions. In its 2007 decision in *Massachusetts v. EPA*, the Supreme Court found that the Environmental Protection Agency has the authority to regulate greenhouse gas emissions from motor vehicles under the Clean Air Act.[4] Further, the court directed EPA to begin the process of determining whether greenhouse gases endanger public health and welfare. If EPA finds that they do, then EPA would be required under the Clean Air Act to regulate their emission. However, it is also possible that EPA would find that greenhouse gases do not endanger public health and welfare, or that there is insufficient evidence to make a finding either way.

Further, in June 2008, the Senate considered legislation (S. 3036) to enact an economy-wide cap-and-trade system to reduce U.S. greenhouse gas emissions.5 However, a cloture motion on this bill failed, and the bill was ultimately tabled. A cap-and-trade system is the favored approach of the incoming Administration, and similar legislation may have a better chance of passage in the 111th Congress.

This report provides background on the evolution of U.S. climate change policy from ratification of the UNFCCC to the George W. Bush Administration's rejection of the Kyoto Protocol programs, to the present. Current major regulatory programs that monitor or reduce greenhouse gas emissions are identified, along with their estimated effect on greenhouse gas emissions. Finally, energy legislation enacted in the 109th and 110th Congresses that could directly or indirectly reduce greenhouse gases is discussed.

Background to Federal Climate Change Policy: From "No Regrets" Back to "No Regrets"

The International Framework

U.S. policy toward global climate change evolved from a "study only" to a "study and action" orientation in 1992 with completion of the UNFCCC in Rio de Janeiro. Both nationally and internationally, much of the debate over policies to address climate change has focused on energy use, because fossil fuel consumption is the main source of greenhouse gas emissions in most countries. During the deliberations on the UNFCCC, the National Academy of Sciences (NAS) released a report on global warming. In this report, *Policy Implications of Greenhouse Warming*, the NAS stated, "The United States could reduce or offset its greenhouse gas emissions by between 10 and 40 percent of 1990 levels at low cost, or at some net savings, if proper policies are implemented."[6] The NAS's energy policy recommendations focused on increasing energy conservation and efficiency, incorporating global warming as a factor in future energy planning, and studying and eventually implementing "full social cost pricing" of energy.

Although the report was widely publicized, many of its recommendations were not applied. Driven by concerns about scientific uncertainty and the potential costs to the economy of measures to reduce greenhouse gas emissions, the George H. W. Bush Administration refused to agree to the negotiation of a binding agreement to reduce the nation's carbon dioxide (CO_2) emissions by a specific date. The UNFCCC reflects the negotiating position of the United States and many other countries in that it called only for voluntary control measures. Senate floor debate on ratification of the UNFCCC brought out concerns by some Senators about the *cost* of compliance, its impact on the country's economic *competitiveness*, and the *comprehensiveness* with respect to the omission of reduction commitments for developing countries—concerns that were lessened because of the non-binding nature of the reduction goals.[7] Those arguing for emissions controls argued that controls could create jobs and enhance economic health, and that high emissions indicated inefficiency.

Asserting that "the developed country Parties should take the lead" in reducing emissions, the UNFCCC set the goal that developed countries aim to return their greenhouse gas emissions to 1990 levels by the year 2000.[8] In line with this goal, developed countries agreed in principle to adopt national plans and policy options to mitigate climate change by reducing anthropogenic emissions and enhancing sinks. The United States submitted such plans in 1992, 1994, 1997, 2002, and 2006, as discussed below.[9]

Developing Programs: EPACT and Climate Action Plans

The Energy Policy Act of 1992 (EPACT) is the principal statutory basis for programs that constitute the U.S. response to the UNFCCC.[10] Programs developed pursuant to EPACT, including appliance energy efficiency standards and updated building codes, are discussed below. Primarily crafted as part of an energy policy response to the Persian Gulf War of 1991, its energy conservation, renewable energy, and other titles were also seen as having a beneficial effect on global climate change concerns being debated at that time in international circles. In its 1992 submission to the UNFCCC, the George H. W. Bush Administration listed 11 different titles of EPACT as "extremely important" to its overall strategy of reducing greenhouse gases.[11]

Some of the previously referenced recommendations of the NAS were embodied in several sections of EPACT. These sections included provisions to establish energy-efficiency standards, promote dissemination of energy-saving information, establish several national research and development programs related to deployment of energy-efficiency technologies, and authorize the Department of Energy (DOE) to evaluate cost-effective energy efficiency technologies. In addition to these activities to improve energy efficiency, EPACT Title XVI aimed to incorporate global warming concerns in energy policy planning. Title XVI authorized DOE to collect, analyze, and report information on climate change. Resulting DOE activities included a report on the various economic, energy, social, environmental, and competitive implications of reducing greenhouse gas emissions; the development of a least-cost energy strategy designed to achieve "the stabilization and eventual reduction in the generation of greenhouse gases"; the creation of a Director of Climate Change; and the development of an inventory of greenhouse gas emissions and early reductions in such emissions.

Indeed, EPACT's authors anticipated that it would help stabilize or even reduce emissions of greenhouse gases at little cost, in line with the 1991 NAS report. As stated by the House report:

> The committee expects that, if fully implemented, H.R. 776 will result in a substantial reduction in U.S. greenhouse gas emissions relative to forecasted levels. The bulk of these reductions result from the programs that will demonstrate and transfer advanced clean coal and renewable technologies abroad, and from the domestic energy efficiency and renewable energy initiatives. The provisions on electric utilities, alternative fuels and coalbed methane are also significant.[12]

The notion that the United States could meet modest CO_2 emission reduction goals at little or no cost underlies many of the global climate change initiatives during the previous Bush and Clinton Administrations, including the George H. W. Bush Administration's "No Regrets" policy and 1992 Climate Action Plan, and the Clinton Administration's 1994 and 1997 Climate Action Plans.[13] Using such an approach to climate change policy, neither of these administrations requested regulatory authority from Congress to implement a climate change policy. Both advocated strategies of undertaking governmental implementing actions that could be done administratively (unless Congress legislated otherwise) and of creating incentives for private industry to voluntarily undertake emissions reduction initiatives.

The Clinton Action Plans were similar to the plan developed under the George H. W. Bush Administration. Both were designed to foster market choices that would conserve energy, increase energy efficiency, and encourage natural gas use. Both were also designed to strengthen selected regulatory standards that concomitantly reduced greenhouse gas emissions—such as landfill regulations that curtail methane releases. Several actions in the 1994 Clinton plan expanded programs listed in the George H. W. Bush Administration's plan by augmenting funding or technical support to increase anticipated reductions. Other Clinton proposals were new: Examples included a "Golden Carrot" program to induce efficiency improvements of industrial equipment; a renewable energy consortium; a program to encourage employers to replace parking subsidies with cash incentives to ride transit, car pool, or find other ways to commute; and a program to promote more efficient nitrogen fertilizer use.[14]

Rejection of the Kyoto Protocol

As it became clear that the voluntary 1992 greenhouse gas emission reduction goals would not be met, parties to the UNFCCC began negotiations that culminated in the 1997 Kyoto Protocol to the UNFCCC. This protocol outlined legally binding emissions reductions for developed countries to specified amounts below 1990 levels, averaged over the years 2008 to 2012. The Clinton Administration committed to a 7% reduction below 1990 levels. The Kyoto Protocol, if it had been submitted to the Senate and ratified, would have changed the U.S. commitment from a voluntary one to a binding commitment. Critics of the Kyoto Protocol raised concerns similar to those debated in connection with the UNFCCC in 1992: concerns about cost, comprehensiveness, and competitiveness. The possibility of failing to comply with a binding commitment intensified the focus on potential costs of the U.S. global climate change policy. The United States, along with most of the world, failed to meet the goal set at Rio of returning 2000 emissions to the level that existed in 1990, a fact that raises questions about the premise that significant greenhouse gas reductions can be achieved at little or no costs.[15] For those who believe substantial reductions in greenhouse gas emissions would entail substantial costs, the Kyoto Protocol's potential costs led to concerns about its effects on the country's competitiveness and its exclusion of developing countries from mandatory emission reductions (comprehensiveness).

That cost, competitiveness, and comprehensiveness remain pivotal factors in climate change policy is illustrated by the George W. Bush Administration's rejection of the Kyoto Protocol early in 2001. In his June 11, 2001, speech on global climate change, the President stated that the Kyoto Protocol was "fatally flawed in fundamental ways." A primary flaw outlined by the President is the exemption of China and other large developing countries from its emissions reduction provisions. This "comprehensiveness" concern was closely followed by "cost" and "competitiveness" concerns. President Bush stated:

> Kyoto is, in many ways, unrealistic. Many countries cannot meet their Kyoto targets. The targets themselves are arbitrary and not based upon science. For America, complying with those mandates would have a negative economic impact with layoffs of workers and price increases for consumers. And when you evaluate all these flaws, most reasonable people will understand that it's not sound public policy.[16]

To respond to global climate change, President Bush called for a new approach focused on science and on flexible control mechanisms that employ market-based incentives. Among the principles that the President argued should guide such a program were the following:

> We must always act to ensure continued economic growth and prosperity for our citizens and for citizens throughout the world.... And finally, our approach must be based on global participation, including that of developing countries whose net greenhouse gas emissions now exceed those in the developed countries.[17]

The Administration's 2001 proposal initiated a new voluntary greenhouse gas reduction program, similar to ones introduced in previous administrations. The plan focuses on improving the carbon intensity of the economy, reducing current emissions of 183 metric tons of carbon equivalent per million dollars of GDP to 151 metric tons per million dollars of GDP by 2012. The plan proposed several voluntary initiatives, along with increased spending and tax incentives, to achieve this goal. However, the Administration stated that three-quarters of the projected reduction would be achieved through current efforts underway, not by new initiatives. The Administration projected that by 2010, the program could result in an emissions reduction of approximately 4.5% relative to "business as usual." However, this level would still be approximately 28% higher than the 1990 level defined by the UNFCCC. Further, without explicit requirements, it is unclear whether the targets set by the Administration will be met.

A key piece of the Administration's proposal was announced on February 12, 2003.[18] Climate, Voluntary Innovative Sector Initiatives: Opportunities Now (Climate VISION) was created in response to President Bush's goal of reducing greenhouse gas intensity of the U.S. economy.[19] Climate VISION aims to assist energy-intensive sectors in developing plans to reduce greenhouse gas intensity, and to publicly recognize the efforts of those sectors.

Another Administration initiative, the Asia-Pacific Partnership on Clean Development and Climate, focuses on international efforts to reduce greenhouse gas emissions outside of the Kyoto Protocol. Of the six countries, the United States and Australia have rejected Kyoto; three countries—China, India, and Korea—are not subject to binding limits under Kyoto. Only one member of the partnership, Japan, has ratified the Kyoto Protocol. The partnership is focused on development and trade of clean energy technologies as well as emissions reductions from key sectors.[20]

This international initiative was followed in May 2007 by the President's announcement that the United States would convene a meeting of the world's "major economies" that are responsible for most greenhouse gas emissions. Held in September 2007, the final statements of the "Major Economies Meeting on Energy Security and Climate Change" emphasized the need to integrate such meetings into the overall UNFCCC negotiations. The U.S. summary of the meeting focused on the "aspirational" nature of reduction goals, reflecting the Administration's rejection of mandatory reduction targets.

REGULATORY PROGRAMS AFFECTING EMISSIONS OF GREENHOUSE GASES

As described above, current federal actions that directly address greenhouse gases focus on research, information, and voluntary programs. Each of the Climate Action Reports submitted by the United States to the UNFCCC has included a compilation of the several dozen programs that various administrations have felt are relevant to reducing U.S. greenhouse gas emissions.[21] Regulatory measures that have reduced greenhouse gas emissions are a small subset of the total U.S. effort numerically, but are responsible for a proportionally larger share of greenhouse gas emission reductions.[22] In general, these regulatory programs were established and implemented primarily for reasons other than climate change concerns. It should be noted that reductions from these programs combined represent about 3% of year 2000 greenhouse gas emissions, and total U.S. emissions have continued to grow: emissions increased approximately 3% between 2000 and 2005; emissions have grown approximately 17% since 1990.[23]

The list of federal regulatory programs discussed here is primarily drawn from activities listed by the George W. Bush Administration in its most recent (2006) submission to UNFCCC.[24] The submission to UNFCCC focused on mandatory programs, but numerous voluntary programs have also been implemented over the past 16 years. This section discusses the seven regulatory programs listed in the Climate Action Report, as well as two additional regulatory programs, carbon dioxide monitoring by electricity generators, and renewable electricity generation requirements (eliminated by P.L. 109-58) not included in the list. These programs' estimated effects on greenhouse gas emissions are summarized in **Table 1**.

Table 1. Estimated Greenhouse Gas Reductions from Federal Regulatory Actions

Program	Estimated GHG Reduction (million metric tons)		
	2002	2012	2020
Landfill Rule	8.7	9.5	9.9
Significant New Alternatives Program	26.0	149.6	222.9
Residential Appliance Standards	N/A	5.1	17.3
Commercial Building Codes	N/A	0.5	3.1
Corporate Average Fuel Economy	0.0	41.8	76.7
Renewable Fuel Standard	N/A	N/A	N/A
Distributed Energy Resources	12.1	23.8	57.2
Total Reductions from Regulatory Programs	47	230	387
Total Reductions from all Federal Programs	255	797	1560

Source: U.S. Department of State, *U.S. Climate Action Report—2006*, Washington, DC, July 2007. Table 4-2.

Note: Totals may not sum due to rounding. "N/A" indicates that although emissions reductions are expected, there may not be enough data currently to estimate those reductions, or that expected reductions are not achieved until later years.

Energy and Environmental Programs Related to Emissions Reductions

Emissions Reductions from Landfills

Section 305 of the 1990 Clean Air Act Amendments requires EPA to control emissions of a variety of air pollutants from new and existing large solid waste landfills.[25] Specifically, the section requires EPA to promulgate New Source Performance Standards (NSPS) for new municipal solid waste landfills, and emissions guidelines for existing landfills to reduce emissions of non-methane organic compounds (NMOCs), including ozone-producing volatile organic compounds (VOCs) and air toxics. Regulations promulgated in 1996 require large landfills that emit landfill gases in excess of 50 metric tons per year to control emissions.[26]

The primary driver for the landfill regulations was reducing formation of ground/surface level ozone (smog), and air toxics. However, in promulgating the rule, the Clinton Administration noted that landfills were the largest U.S. source of emissions of the greenhouse gas methane (40%), and that the rule would have the indirect benefit of reducing methane emissions by 50%.[27] In its 2006 Climate Action Report, the current Administration estimated that the year 2002 methane emissions reductions achieved by the rule were 8.7 million metric tons of carbon dioxide equivalent, predicted to increase to 9.9 million metric tons by 2020.[28]

Significant New Alternatives Policy (SNAP) Determinations

Many ozone-depleting substances are also greenhouse gases (such as perfluorocarbons). Therefore, efforts to protect the ozone layer also tend to reduce greenhouse gas emissions. There is a complex scientific relationship between ozone depletion, ozone-depleting chemicals and climate change.

Title VI of the 1990 Clean Air Act Amendments represents the United States' primary response on the domestic front to the ozone depletion issue. It also implements this country's international responsibilities under the Montreal Protocol to Reduce Ozone-Depleting Substances.[29] Section 612 requires EPA to develop a program to identify alternatives to ozone depleting substances banned under the Montreal Protocol.

In determining the acceptability of an alternative, EPA is to assess the overall risk to human health or the environment that the alternative poses, compared with other alternatives. In promulgating the implementing regulation for the program in 1994, EPA identified increased global warming as one of the risk criteria that it would use in determining the acceptability of an alternative.[30]

SNAP determinations focus on the global warming potential of various substitutes used in place of the ozone-depleting chemicals banned under the Montreal Protocol, not on the global warming potential of the banned ozone-depleting chemicals themselves.

Under the regulation, EPA has restricted or narrowed the use of hydrofluorocarbons (HFCs) and perfluorocarbons (PFCs) where alternatives with lower global warming potentials exist. EPA estimates that the restrictions reduced greenhouse gas emissions by 26 million metric tons of carbon dioxide equivalent in 2002, and projects a further reduction of nearly 200 million metric tons (for a total of 223 million metric tons) by 2020.[31]

Residential Appliance Standards

The 1987 National Appliance Energy Conservation Act (NAECA)[32] set minimum efficiency standards for many major appliances.[33] The Energy Policy Act of 1992 (EPACT) expanded the list of covered appliances and allowed for future standards development for other products.[34] Under NAECA and EPACT, the Department of Energy must develop mandatory energy efficiency standards for these appliances, and review them in accordance with a statutorily set schedule to determine whether they are sufficiently stringent. DOE is required to set standards designed to achieve the maximum improvement in energy efficiency it believes is "technically feasible and economically justifiable."[35]

The primary driver for residential appliance standards has been energy conservation. In 1997, the Clinton Administration estimated that the appliance standards would save almost a quad (1 quadrillion Btu) of energy, resulting in a 21.6 million metric ton reduction in carbon emissions by the year 2010.[36] For the 2006 Climate Action Report, DOE estimates reductions of 5 million metric tons of carbon dioxide in 2012, increasing to 17 million metric tons in 2020.[37]

Updating State Commercial Building Codes

Section 101 of EPACT requires DOE to review and update provisions of the commercial building code with respect to energy efficiency.[38] Specifically, DOE is directed to determine whether revisions to the ASHRAE[39] Standard 90.1—1989 (and any future revisions) would improve energy efficiency in commercial buildings. If DOE determines that revisions would improve energy efficiency, states are required to review and update their commercial building codes accordingly, with respect to improving energy efficiency. In

July 2002, DOE determined that ASHRAE/IESNA Standard 90.1—1999 would save energy in commercial buildings.[40] Thus, states were required to review and update their commercial building codes by July 15, 2004.[41] The Energy Policy Act of 2005 requires further updates to state commercial building standards.

In announcing its July 2002 determination, the DOE focused on the energy savings that state adoption of the standard would entail, estimated at 130 million barrels of oil equivalent over 10 years.[42] The DOE announcement does not mention any carbon dioxide reduction that could result from the improved energy efficiency. In the 2006 Climate Action Plan, DOE estimates that the new requirements will save 0.5 million metric tons of carbon dioxide in 2012, increasing to 3.1 million metric tons in 2020.[43]

Corporate Average Fuel Economy (CAFE)

Federal fuel economy standards directly affect greenhouse gas emissions from the transportation sector. The Energy Policy and Conservation Act of 1975 (EPCA) established corporate average fuel economy (CAFE) standards for new passenger cars, and gave the Department of Transportation (DOT) the authority to set standards for other vehicles, including "light trucks," which consist of pickups, vans, and sport utility vehicles .[44] The goal of EPCA was to reduce the dependence on foreign oil after the Arab oil embargo of the 1970s.

Increasing CAFE standards would lead to reductions in fuel consumption and greenhouse gas emissions as older, less efficient vehicles were replaced with more efficient models. However, concerns associated with increased CAFE standards include questions of occupant safety and vehicle choice. In 1994, the Clinton Administration considered raising the CAFE standard for light trucks. However, Congress included language in the annual FY1996-FY2001 DOT Appropriations Acts[45] prohibiting the use of appropriated funds for any rulemaking on CAFE, effectively freezing the standards. However, the Senate conferees on the FY2001 appropriations insisted upon a study of CAFE by the National Academy of Sciences (NAS). That study, released on July 30, 2001, concluded that it would be possible to achieve a more than 40% improvement in light truck fuel economy over a 10- to 15-year period at costs that would be recoverable over the lifetime of vehicle ownership, without compromising safety.

On April 7, 2003, DOT announced a final rule increasing light truck CAFE standards to 22.2 miles per gallon by model year 2007.[46] For the 2006 Climate Action Report, DOT estimated that in total, the regulations would save 42 million metric tons of carbon dioxide in 2012.[47]

On April 6, 2006, DOT announced further changes to the light truck CAFE standards starting in model year 2008.[48] By model year 2011, all light truck manufacturers will be subject to new standards based on a vehicle's size. DOT estimates that the new rule will save an additional 4.4 billion gallons of gasoline over the life of the vehicles produced between model years 2008 and 2011 (relative to the MY2007 standard). This would mean an additional greenhouse gas reduction of roughly 10 million to 12 million metric tons. Both rulemakings combined would lead to an estimated reduction of between 18 million and 22 million metric tons of greenhouse gases.

However, on November 15, 2007, the U.S. Court of Appeals for the Ninth Circuit remanded the MY2008 standards back to DOT because the agency did not explicitly value greenhouse gas reductions in its estimates of the costs and benefits of the rulemaking.[49]

On December 19, 2007, President Bush signed the Energy Independence and Security Act of 2007 (EISA, P.L. 110-140). Among other provisions, the new law requires a significant increase in combined passenger car/light truck fuel economy. By 2020, the combined new vehicle fleet must meet a combined CAFE average of 35 mpg, up from roughly 25 mpg in 2007. This fuel economy increase could lead to a significant reduction in greenhouse gas emissions from projected levels. In its Preliminary Regulatory Impact Analysis on proposed regulations for Model Years 2011-2015, DOT estimates that proposed rules for those years would save 521 million metric tons of CO_2 over the life of those vehicles.[50] The American Council for an Energy-Efficient Economy estimates that the new CAFE standards could save roughly 400 million metric tons per year and 3,800 million metric tons total by 2030.[51] EISA is discussed in greater detail below.

Renewable Fuel Standard

The Energy Policy Act of 2005 established a renewable fuel standard (RFS) requiring the use of renewable fuels in gasoline. EISA further expanded this requirement, and for the first time set requirements for the lifecycle greenhouse gas emissions of motor fuels. For the 2006 Climate Action Report, EPA did not estimate emissions reductions from the program. However, it is likely that the program could lead to lower emissions, especially with the new requirements of EISA.

Distributed Energy Resources

A final program that the George W. Bush Administration describes as regulatory is the Distributed Energy Program at the Department of Energy.

While most of the program is focused on R&D and commercialization of new technologies, the program also aims to eliminate regulatory barriers to the use of distributed energy. For all facets of the program, DOE estimates that the program saved 12.1 million metric tons of carbon dioxide in 2002, and will save 57 million metric tons in 2020.[52]

Monitoring Rules—Carbon Dioxide Monitoring by Electric Generating Facilities

Section 821 of the 1990 Clean Air Act Amendments requires electric generating facilities affected by the acid rain provisions of Title IV to monitor carbon dioxide in accordance with Environmental Protection Agency (EPA) regulations.[53] This provision was enacted for the stated purpose of establishing a national carbon dioxide monitoring system.[54] As promulgated by EPA, regulations permit owners and operators of affected facilities to monitor their carbon dioxide emissions through either continuous emission monitoring (CEM) or fuel analysis.[55] The CEM regulations for carbon dioxide are similar to those for the acid rain program's sulfur dioxide CEM regulations. Those choosing fuel analysis must calculate mass emissions on a daily, quarterly, and annual basis, based on amounts and types of fuel used.

Although this regulation does not actually reduce carbon dioxide emissions, it does expressly target the global climate change issue. Also, it represents a necessary first step toward any future market-oriented greenhouse gas reduction program. Whether a market-oriented control program were to be based on tradable emissions credits or a carbon tax, accurate emissions data would be the foundation for developing the allocation systems, reduction targets, and enforcement provisions.

Regulatory Program Promoting Renewable Energy through PURPA Eliminated by P.L. 109-58

The 1978 Public Utility Regulatory Policies Act (PURPA) is designed to encourage the development of cogeneration and small power production (called "qualifying facilities" or QFs).[56] In particular, Section 210 contained a mandatory purchase clause requiring utilities to buy power from QFs at the utilities' avoided cost. PURPA exempted from the full breadth of federal and state regulation certain small power producers, including those using geothermal, solar, wind, and waste energy.[57] This regulatory exemption, along with the mandatory purchase requirement contained in Section 210, has proven to be a strong incentive for development of renewable energy, particularly biomass. QF renewable capacity represents a substantial majority of U.S. non-hydroelectric renewable energy capacity.[58] The Energy Policy Act of 2005 repealed the Section 210 purchase requirement for new contracts (see the section below on "The Energy Policy Act of 2005").

PURPA was enacted to promote energy security through energy conservation and the development of alternative energy sources in the aftermath of the 1974 oil crisis. As events have unfolded, the effort to reduce dependence on fossil fuels has had the additional perceived benefit of reducing carbon dioxide emissions. In 1997, the Clinton Administration estimated that its renewable energy commercialization efforts with respect to biomass, geothermal, and wind would reduce greenhouse gas emissions by 17.6 million metric tons of carbon equivalent by the year 2010.[59] For the 2002 Climate Action Report, the George W. Bush Administration provided no specific estimate of reductions from the use of renewable electricity. However, for the general category of energy supply, the Administration estimated the year 2000 effect to have been a saving of 14.7 million metric tons of carbon dioxide equivalent.[60]

With the passage of the Energy Policy Act of 2005, Section 210 was amended, subject to a FERC determination of market conditions. In October 2006, FERC adopted a final rule delineating those conditions, finding that utilities operating within regional transmission organizations (RTOs) met the conditions for repeal, while the market conditions for non-RTO utilities would be determined on a case-by-case basis.[61]

THE ENERGY POLICY ACT OF 2005

On August 8, 2005, President Bush signed the Energy Policy Act of 2005 (P.L. 109-58), with provisions directly and indirectly related to greenhouse gas emissions. Title XVI establishes a voluntary national program designed to encourage voluntary reductions in greenhouse gases. The effort is led by an Interagency Committee, with DOE playing a key supporting role. Title XVI attempts to support actions focused on reducing U.S. carbon intensity, but does not establish a requirement to reduce emissions. The title also establishes a program to encourage exports of carbon intensity-reducing technologies to developing countries. This program is led by the Secretary of State.

In addition to Title XVI, Section 1253 repeals the mandatory purchase requirement under Section 210 of PURPA for new contracts if FERC finds that a competitive electricity market exists and if other conditions are met.[62] The debate over the bill included proposals to increase CAFE standards and to establish a renewable portfolio standard, although these changes were not included in the final law.

Also not included in P.L. 109-58 was Section 1612 of the Senate bill (H.R. 6, as amended by S.Amdt. 866), which expressed the Sense of the Senate that human activities are a substantial cause of greenhouse gas accumulation in the atmosphere, causing average temperatures to rise. Further, the resolution called for a mandatory, market-based program to limit greenhouse gas emissions. Also, a bill to establish a mandatory, market-based greenhouse gas reduction program (S. 1151) was debated on the Senate floor as S.Amdt. 826 and defeated by a 38-60 vote.

THE ENERGY INDEPENDENCE AND SECURITY ACT OF 2007

On December 19, 2007, President Bush signed the Energy Independence and Security Act of 2007 (EISA, P.L. 110-140). EISA contains many energy provisions that could lead to reductions in greenhouse gas emissions.[63] In addition to these indirect reductions, EISA also directly addresses climate change issues in several ways.

First, EISA expands the renewable fuel standard (RFS) established in P.L. 109-58. The RFS requires that a minimum amount of renewable fuels be blended into transportation fuels each year. The EISA amendments to the RFS significantly expand the mandated level. Further, they require that an increasing share of the RFS be met with "advanced biofuels" defined as having 50% lower lifecycle greenhouse gas emissions than petroleum fuels. This is the first time that Congress has enacted national policy addressing the carbon content of motor fuels.

Second, Title VII of the new law focuses on research, development, and demonstration of technologies to capture and store carbon dioxide. DOE research and development is expanded and will include large-scale demonstration projects. The Department of the Interior must develop a methodology to assess the national potential for geologic and ecosystem storage of carbon dioxide, and must recommended a regulatory framework for managing geologic carbon sequestration on public lands.

In addition to the above programs, EISA also requires the establishment of an Office of Climate Change and Environment in the Department of Transportation. This office will plan, coordinate, and implement research at DOT on reducing transportation-related energy use, mitigating the causes of climate change, and addressing the impacts of climate change on transportation.

Energy provisions not directly addressing climate change, but that could lead to lower greenhouse gas emissions, include

- more stringent fuel economy (CAFE) standards for passenger cars and light trucks;
- higher-efficiency standards for appliances and lighting;
- higher-efficiency requirements for government buildings; and
- research and development on renewable energy.

Table 2. 2005 Energy-Related Greenhouse Gas Emissions (million metric tons of CO_2 equivalent)

Greenhouse gas	Energy-related 2005 emissions	Total U.S. 2005 emissions	Energy as percent of total
Carbon Dioxide[a]	5,943	6,090	98%
Methane	206	539	38%
Nitrous Oxide	52	469	11%
HFC, PFC, SF6	13	163	8%
Totals[b]	6,215	7,260	86%

Source: Environmental Protection Agency, *Inventory of U.S. Greenhouse Gas Emissions and Sinks: 1990-2005* April 2007

Note: Because carbon dioxide is by far the largest volume greenhouse gas, energy-related emissions are responsible for the majority of total greenhouse gas emissions.

a. Biomass not included.

b. Totals may not add due to independent rounding.

CONCLUSION

Energy policy was a key issue in the 110th Congress, as it was in earlier Congresses. High energy prices and international instability motivated passage of the 2005 Energy Policy Act and the 2007 Energy Independence and Security Act. Given that energy consumption is the dominant source of carbon dioxide emissions in this country, and a substantial source of overall greenhouse gas emissions, any reduction in energy consumption will likely lead to lower emissions. As indicated below (**Table 2**), energy-related activities are responsible for about 86% of the country's greenhouse gas emissions, and 98% of its carbon dioxide emissions.

As noted in this report, climate change was an integral part of the debate on the 1992 Energy Policy Act (EPACT), occurring as it did during the signing and ratifying of the UNFCCC.[64] Indeed, EPACT became the implementing legislation for the UNFCCC, and, as discussed in this report, those programs are responsible for much of the reduced growth in greenhouse gases achieved since 1992. Most federal policies and regulations that have resulted in greenhouse gas reductions were, in fact, promulgated as energy policy initiatives. However, these policies have not reversed growth in overall U.S. emissions.

Climate change—as a specific issue needing a regulatory response—was debated during deliberations on the Energy Policy Act of 2005 (EPAct 2005) and recognized in passage of Senate Amendment 866 expressing the Sense of the Senate that Congress should enact a comprehensive and effective national program of mandatory, market-based limits and incentives on greenhouse gases that slow, stop, and reverse the growth of such emissions. The Energy Independence and Security Act (EISA) of 2007 directly addresses some climate change issues, but still generally focuses on energy supply and consumption. The overall effect of EPAct 2005 and EISA on future greenhouse gas emissions is unclear, and specific action on the issue will depend on efforts to enact the national program called for in Senate Amendment 866. An attempt to pass an economy-wide greenhouse gas cap-and-trade program in the Senate was unsuccessful in the 110th Congress.

While some provisions in energy laws enacted over the past 16 years have led to lower greenhouse gas emissions or addressed climate change directly, other provisions in those same laws have almost certainly resulted in higher emissions.[65] To date, no energy law has had reducing greenhouse gas emissions as the main organizing principle. Comprehensive climate change policy directed at reducing greenhouse gas emissions should address energy supply and consumption and, thus, be integrated with energy policy. This will be a pivotal challenge to the 111th Congress's and the incoming Administration's anticipated efforts to enact legislation to limit greenhouse gas emissions.

End Notes

[1] For further information, see CRS Report RL33826, *Climate Change: The Kyoto Protocol, Bali "Action Plan," and International Actions*, by Susan R. Fletcher and Larry Parker.

[2] Most of the programs outlined in the report involve research, technical assistance, information gathering, or technical assistance programs initiated by the federal government, or voluntary emissions reduction programs coordinated by the government.

[3] The "no regrets" policy was one of establishing programs for other purposes, that would have concomitant greenhouse gas reductions. Therefore, only those policies that reduced greenhouse gas emissions at no additional cost were considered.

[4] For more information, see CRS Report RS22665, *The Supreme Court's Climate Change Decision: Massachusetts v. EPA*, by Robert Meltz.

[5] For more information on cap-and-trade legislation in the 110th Congress, see CRS Report RL33 846, *Greenhouse Gas Reduction: Cap-and-Trade Bills in the 110th Congress*, by Larry Parker, Brent D. Yacobucci, and Jonathan L. Ramseur.

[6] National Academy of Sciences, *Policy Implications of Greenhouse Warming* (Washington, DC: National Academy Press, 1991), p. 73.

[7] *Congressional Record*, vol. 138 (October 7, 1992), 33520-33527.

[8] *United Nations Framework Convention on Climate Change (UNFCCC)*, Article 3, Section 1.

[9] For a more detailed discussion, see CRS Report RL30024, *U.S. Global Climate Change Policy: Evolving Views on Cost, Competitiveness, and Comprehensiveness*, by Larry Parker and John Blodgett.

[10] The other primary source of greenhouse-gas related regulations is the Clean Air Act, particularly the 1990 Amendments. Clean Air Act regulations—concerning mandatory carbon dioxide monitoring by electricity generators, landfill emissions reductions, and the control of ozone depleting substances—are discussed later in this report.

[11] Department of State, *National Action Plan for Global Climate Change* (Washington, DC: Department of State, 1992), p. 73.

[12] Committee on Energy and Commerce, *Comprehensive National Energy Policy Act*, H.Rept. 102-474, Part 1, March 30, 1992, p. 152.

[13] On the "no regrets" policy of the George H. W. Bush Administration, see C. Boyden Gray and David B. Rivkin, Jr., "A 'No Regrets' Environmental Policy," *Foreign Policy*, summer 1991, pp. 47-65; for the various action plans, see U.S. Department of State, *National Action Plan for Global Climate Change*, Department of State Publication 10026, December 1992; U.S. Department of State, *Climate Action Report*, Department of State Publication, 1994; and U.S. Department of State, *Climate Action Report*, Department of State Publication 10496, July 1997.

[14] President Clinton and Vice President Gore, *The Climate Change Action Plan*, White House, October 1993.

[15] Indeed, U.S. emissions of greenhouse gases were 14% above 1990 levels in 2000. Environmental Protection Agency, *Inventory of U.S. Greenhouse Gas Emissions and Sinks*, Washington, DC, April 2002, p. ES-3.

[16] President George W. Bush, *President Bush's Speech on Global Climate Change*, June 11, 2001.

[17] Ibid.

[18] Environmental Protection Agency, *Bush Administration Launches "Climate VISION."* February 12, 2003.

[19] Greenhouse gas intensity is a measure of emissions per unit of economic activity (often expressed as tons of emissions per thousand or million dollars of Gross Domestic Product). For more on greenhouse gas intensity, see CRS Report 98-23 5, *Global Climate Change: U.S. Greenhouse Gas Emissions—Status, Trends, and Projections*, by John Blodgett and Larry Parker.

[20] http://www.asiapacificpartnership.org/

[21] For the most recent compilation, see U.S. Department of State, *U.S. Climate Action Report—2006*, Washington, DC, July 2007.

[22] For example, EPA regulations on landfills and Significant New Alternatives Policy Determinations (SNAP) account for 35 million metric tons of the 255 million metric tons (carbon dioxide equivalent) in claimed reductions by the George W. Bush Administration for the year 2002 (14% of total). U.S. Department of State, *U.S. Climate Action Report*—2006, Washington, DC, July 2007. Both the landfill regulations and SNAP are discussed below.

[23] U.S. Energy Information Administration, *Emissions of Greenhouse Gases in the United States 2005*, Washington, DC, November 2006.

[24] Ibid.

[25] Landfill gas contains methane, carbon dioxide, and numerous non-methane organic compounds (NMOCs), including vinyl chloride, toluene, and benzene.

[26] 61 *Federal Register* 9905-9944 (March 12, 1996).

[27] EPA, *Fact Sheet: Final Air Regulations for Municipal Solid Waste Landfills*, March 1, 1996. This is due to the fact that techniques to reduce methane and NMOCs are basically the same.

[28] U.S. Department of State, *U.S. Climate Action Report—2006*, Washington, DC, July 2007, p. 59.

[29] It should be noted that emissions controlled under the Montreal Protocol are separate from those controlled by the Kyoto Protocol, and any reductions in ozone-depleting compounds would not count toward Kyoto Protocol targets.

[30] 59 *Federal Register* 13049 (March 18, 1994).

[31] U.S. Department of State, *U.S. Climate Action Report—2006*, Washington, DC, July 2007, p. 58.

[32] P.L. 100-12.

[33] As amended in 1988, appliances included refrigerators, refrigerator-freezers, freezers, room air conditioners, fluorescent lamp ballasts, incandescent reflector lamps, clothes dryers, clothes washers, dishwashers, kitchen ranges and ovens, pool heaters, television sets (withdrawn in 1995), and water heaters.

[34] Additional appliances included commercial building heating and air conditioning equipment, water heaters, certain incandescent and fluorescent lamps, distribution transformers, and electric motors.

[35] 10 CFR Chapter II, Part 430. For a summary of resulting standards, see http://www.eere.energy appliance_standards/residential_products.html.

[36] U.S. Department of State, *U.S. Climate Action Report 1997*, Washington, DC, July 1997, p. A-7.

[37] U.S. Department of State, *U.S. Climate Action Report—2006*, Washington, DC, July 2007, p. 55.

[38] Section 101, *Energy Policy Act of 1992*, P.L. 102-486. The section contains provisions for updating residential building codes; however, those provisions only require states to review any DOE determination with respect to updating codes. Adoption of the revisions is not mandatory.

[39] American Society of Heating, Refrigeration, and Air Conditioning Engineers.

[40] 67 *Federal Register* 46464 (July 15, 2002).

[41] As of January 2007, 10 states and had not updated their commercial building energy codes to comply with the EPACT 1992, according to the Building Codes Assistance Project. However, in three of those states, the codes have been adopted in some jurisdictions. See http://www.bcap-energy.org. The Energy Policy Act of 2005 requires a further update to the standards. An additional five states and the District of Columbia have yet to update their codes to the new requirement.

[42] DOE, *Energy Department Determines that Model Commercial Building Code Will Save Energy & Benefit Consumer*, July 15, 2002. Available at http://www.energycodes.gov/implement/determinations_com_news.stm.

[43] U.S. Department of State, *U.S. Climate Action Report—2006*, Washington, DC, July 2007, p. 55.

[44] P.L. 94-163.

[45] P.L. 104-50 (FY1996), P.L. 104-205 (FY1997), P.L. 105-66 (FY1998), P.L. 105-277 (FY1999), P.L. 106-69 (FY2000), and P.L. 106-346 (FY2001).

[46] 68 *Federal Register* 16867-16900. April 7, 2003.

[47] U.S. Department of State, *U.S. Climate Action Report—2006*, Washington, DC, July 2007, p. 57.

[48] 71 *Federal Register* 17655-17679. April 6, 2006.

[49] United States Court of Appeals, Ninth Circuit. *Center for Biological Diversity vs. National Highway Traffic Safety Administration*. Argued and Submitted August 14, 2007. Filed November 15, 2007.

[50] National Highway Traffic Safety Administration, *Preliminary Regulatory Impact Analysis: Corporate Average Fuel Economy for MY2011-2015 Passenger Cars and Light Trucks*, Washington, DC, April 2008, pp. III-17, http://www.nhtsa.gov/staticfiles/DOT

[51] American Council for an Energy-Efficient Economy, *Energy Bill Savings Estimates as Passed by the Senate*, Washington, DC, December 14, 2007, http://www.aceee.org/energy

[52] U.S. Department of State, *U.S. Climate Action Report—2006*, Washington, DC, July 2007, p. 56.

[53] Section 821, *1990 Clean Air Act Amendments* (P.L. 101-549, 42 USC 7651k).

[54] S.Rept. 101-952.

[55] See 40 CFR 75.13, along with appendix G (for CEMs specifications) and appendix F (for fuel analysis specifications).

[56] P.L. 95-617 (1978).

[57] As originally enacted, the law limited most small power producers to 30 MW, with geothermal energy limited to 80 MW. The Solar, Wind, Waste, and Geothermal Production Incentives Act of 1990 (P.L. 101-575) removed the size limitations.

[58] Energy Information Administration, *Renewable Energy 2000: Issues and Trends*, Washington, DC, February 2001, p. 10.

[59] U.S. Department of State, *U.S. Climate Action Report 1997*, Washington, DC, July 1997, pp. A-23, A-24, and A-27.

[60] U.S. Department of State, *U.S. Climate Action Report 2002*, Washington, DC, May 2002, p. 65.

[61] This determination represents only a preliminary, rebuttable, presumption that conditions have been met. Utilities are still required to file for relief from Section 210 requirements. For more information, see FERC docket RM06-10-000, October 19, 2006.

[62] Proponents of renewable energy argued that repeal would reduce the role of renewables, while advocates of repeal argued that renewables no longer need economic intervention. For a more detailed discussion of PURPA, see CRS Report RL32728, *Electric Utility Regulatory Reform: Issues for the 109th Congress*, by Amy Abel.

[63] For more information on EISA, see CRS Report RL34294, *Energy Independence and Security Act of 2007: A Summary of Major Provisions*, coordinated by Fred Sissine.

[64] For more on the relationship between UNFCCC and EPACT, see CRS Report RL3 0024, *U.S. Global Climate Change Policy: Evolving Views on Cost, Competitiveness, and Comprehensiveness*, by Larry Parker and John Blodgett.

[65] These include provisions that promote additional production and/or use of fossil fuels.

In: Climate Change Litigation and Law
Editors: Jean-François Masson ISBN: 978-1-60876-089-3

Chapter 4

HEARING ON IMPLICATIONS OF SUPREME COURT'S EPA DECISION-BROWNER TESTIMONY

Carol M. Browner
Committee on Environment and Public Works
United States Senate

Good morning, Madam Chairman, Senator Inhofe, and members of the Committee. I appreciate the opportunity to speak to you today on the most pressing environmental and public health issue that our country, and for that matter the world, has ever faced. That is, climate change.

During my eight years as the Administrator of the EPA, I worked hard to protect the environment and public health, both for our generation and future generations. Today, I would like to discuss how the current EPA can use the mandate given by the Supreme Court in Massachusetts v. EPA to respond to the climate change crisis immediately.

First — the EPA can grant California its federal waiver to enforce its own greenhouse gas standards. California has thus far outpaced the federal government on greenhouse gas regulation — it has ignored critics and naysayers, moving ahead with an aggressive plan. The EPA should in turn recognize this plan by granting California the authority to put it into place. Second — EPA should act now on setting greenhouse gas standards for vehicles and power plants, two significant sources of emissions.

These are a few things EPA can do right now to regulate emissions, but it is not enough. The magnitude of the Supreme Court decision warrants Congressional leadership and immediate action as well.

As we seek to address climate change, both through the actions of EPA and through Congress, three realities should guide us. First, that the science on climate change cannot, at this point, be in doubt. Second, that we can find common-sense, cost-effective ways to regulate greenhouse gas emissions. And third, that EPA and Congress now have the undisputed authority and responsibility to regulate the emission of greenhouse gases.

When it comes to science, the facts continue to roll in, and the scientific community has reached a consensus. The considered judgment of twenty-five hundred of the world's top climate change scientists, 11 national scientific academies, and hundreds of scientists contributing to the IPCC is simply this: climate change is real, it is caused by human activities, it is rapidly getting worse, and it will transform both our planet and humanity if action is not taken now.

Such action need not bankrupt us or disrupt our economy. We can and we must find cost- effective ways to meet greenhouse gas standards. Historically, American innovation and ingenuity have served us well. Let us harness them now. In the past, we have been willing to set standards without having in hand the actual technology necessary to meet such standards. For example, when Congress decided to ban chlorofluorocarbons, there was no technology to replace CFC's. But once Congress made the decision, there was a guaranteed market for replacement; companies competed with each other and, within a relatively short time, there was a replacement, and at far less cost than had been anticipated. We may not have a perfect formula for cutting greenhouse gas emissions yet, but that is no reason to hold off on setting regulations and enforcing them.

The EPA has the moral — and now the legal — authority to set greenhouse standards in accordance with the Clean Air Act to limit climate change and protect the heath of future generations. It is said that nine-tenths of wisdom is being wise in time. Congress also has the prerogative to ensure that EPA does its duty and to take bold action on its own.

We have the science; the will has been summoned; the technology will come. Have no doubt - we can stop global warming. Anything less would be a felony against the future, a failure to meet our responsibility to our children and theirs. My request is that we do our duty.

Thank you very much. Now I would be pleased to respond to any questions you might have.

Thank you.

Attachment A: "EPA's Authority to Regulate Pollutants Emitted by Electric Power Generation Sources" Memorandum from Jonathan Z. Cannon to Carol M. Browner, April 10, 1998

Attachment B: Testimony of Gary S. Guzy before a joint hearing of the Subcommittee on National Economic Growth, Natural Resources and Regulatory Affairs of the Committee on Government Reform and the Subcommittee on Energy and Environment of the Committee on Science, US House of Representatives, October 6, 1999

ATTACHMENT A

United States Environmental Protection Agency
Washington, D.C. 20460
Office of General Counsel

APR 10 1998

MEMORANDUM

SUBJECT: EPA's Authority to Regulate Pollutants Emitted by Electric Power Generation Sources

FROM: Jonathan Z. Cannon
General Counsel

TO: Carol M. Browner
Administrator

I. Introduction and Background

This opinion was prepared in response to a request from Congressman DeLay to you on March 11, 1998, made in the course of a Fiscal Year 1999 House Appropriations Committee Hearing. In the Hearing, Congressman DeLay referred to an EPA document entitled "Electricity Restructuring and the Environment: What Authority Does EPA Have and What Does It Need."

Congressman DeLay read several sentences from the document stating that EPA currently has authority under the Clean Air Act (Act) to establish pollution control requirements for four pollutants of concern from electric power generation: nitrogen oxides (NOx), sulfur dioxide (SO2), carbon dioxide (CO2), and mercury. He also asked whether you agreed with the statement, and in particular, whether you thought that the Clean Air Act allows EPA to regulate emissions of carbon dioxide. You agreed with the statement that the Clean Air Act grants EPA broad authority to address certain pollutants, including those listed, and agreed to Congressman DeLay's request for a legal opinion on this point. This opinion discusses EPA's authority to address all four of the pollutants at issue in the colloquy, and in particular, CO2, which was the subject of Congressman DeLay's specific question.

The question of EPA's legal authority arose initially in the context of potential legislation addressing the restructuring of the utility industry. Electric power generation is a significant source of air pollution, including the four pollutants addressed here. On March 25, 1998, the Administration announced a Comprehensive Electricity Plan (Plan) to produce lower prices, a cleaner environment, increased innovation and government savings. This Plan includes a proposal to clarify EPA's authority regarding the establishment of a cost-effective interstate cap and trading system for NOx reductions addressing the regional transport contributions needed to attain and maintain the Primary National Ambient Air Quality Standards (NAAQS) for ozone. The Plan does not ask Congress for authority to establish a cap and trading system for emissions of carbon dioxide from utilities as part of the Administration's electricity restructuring proposal. The President has called for cap-and-trade authority for greenhouse gases to be in place by 2008, and the Plan states that the Administration will consider in consultation with Congress the legislative vehicle most appropriate for that purpose.

As this opinion discusses, the Clean Air Act provides EPA authority to address air pollution, and a number of specific provisions of the Act are potentially applicable to control these pollutants from electric power generation. However, as was made clear in the document from which Congressman DeLay quoted, these potentially applicable provisions do nor easily lend themselves to establishing market-based national or regional cap-and-trade programs, which the Administration favors for addressing these kinds of pollution problems.

IL Clean Air Act Authority

The Clean Air Act provides that EPA may regulate a substance if it is (a) an "air pollutant,"and (b) the administrator makes certain findings regarding such pollutant (usually related to danger to public health, welfare, or the environment) under one or more of the Act's regulatory provisions.

A. Definition of Air Pollutant

Each of the four substances of concern as emitted from electric power generating units falls within the definition of "air pollutant. under section 302(g). Section 302(g) defines air pollutant" as

> any air pollution agent or combination of such agents, including any physical, chemical, biological, [or] -radioactive . substance or matter which is emitted into or otherwise enters the ambient air. Such term includes any precursors to the formation of any air pollutant, to the extent that the Administrator has identified such precursor or precursors for the particular purpose for which the term "air pollutant" is used.

This broad definition states that "air pollutant" includes any physical, chemical, biological, or radioactive substance or matter that is emitted onto or otherwise enters the ambient air SO2, NOx, CO2, and mercury from electric power generation are each a "physical [and] chemical... substance which is emitted into .. the ambient air," and hence, each is an air pollutant within the meaning of the Clean Air Act.[1]

A substance can be an air pollutant even though it is naturally present in air in some quantities. Indeed, many of the pollutants that EPA currently regulates are naturally present in the air in some quantity and are emitted from natural as well as anthropogenic sources. For example, SO2 is emitted from geothermal sources; volatile organic compounds (precursors to ozone) are emitted by vegetation and particulate mater and NOx, are formed from natural sources through natural processes, such a naturally occurring forest fires. Some substances regulated under the Act as hazardous air pollutants are actually necessary in trace quantities for human life, but are toxic at higher levels or through other routes of exposure. Manganese and selenium are two examples of such pollutants. EPA regulates a number of naturally occurring substances as air pollutants, however, because human activities have increased the quantities present in the air to levels that are harmful to public health, welfare, or the environment.

B. EPA Authority to Regulate Air Pollutants

EPA's regulatory authority extends to air pollutants, which, as discussed above, are defined broadly under the Act and include SO2, NOx, CO2, and mercury emitted into the ambient air. Such a general statement of authority is distinct from an EPA determination that a particular air pollutant meets the specific criteria for EPA action under a particular provision of the Act. A number of specific provisions of the Act are potentially applicable to these pollutants emitted from electric power generation.[2] Many of these specific provisions for EPA action share a common feature in that the exercise of EPA's authority to regulate air pollutants is linked to determination by the Administrator regarding the air pollutants' actual or potential harmful effects on public health, welfare or the environment. See also sections 108, 109, 111(b), 112, and 115. See also sections 202(a), 211(c), 231, 612, and 615. The legislative history of the 1977 Clean Air Act Amendments provides extensive discussion of Congress' purposes in adopting the language used throughout the Act referencing a reasonable anticipation that a substance endangers public health or welfare. One of these purposes was "to emphasize the preventative or precautionary nature of the act, i.e., to assure that regulatory action can effectively prevent harm before it occurs, to emphasize the predominant value of protection of public health." H.R. Rep. No. 95294 95th Cong., 1st Sess, at 49 (Report of the Committee on Interstate and Foreign Commerce). Another purpose was "No assure that the health of susceptible individuals, as well as healthy adults, will be encompassed in the term 'public health,'..." Id. at 50. "Welfare" is defined in section 302(h) of the Act, which states:

> [a]lI language referring to effects on welfare includes, but is not limited to, effects on soils, water, crops, vegetation, man-made materials, animals, wildlife, weather, visibility, and climate, damage to and deterioration of property, and hazards to transportation, as well as effects on economic values and on personal comfort and well-being, whether caused by transformation, conversion, or combination with other air pollutants.[3]

EPA has already regulated SO2, NOx, and mercury based on determinations by EPA or Congress that these substances have negative effects on public health, welfare, or the environment. While CO2, as an air pollutant, is within EPA's scope of authority to regulate, the Administrator has not yet determined that CO2 meets the criteria for regulation under one or more provisions of the Act. Specific regulatory criteria under various provisions of the Act could be met if the Administrator determined under one or more of those provisions that CO2 emissions are reasonably anticipated to cause or contribute to adverse effects on public health, welfare, or the environment.

C. EPA Authority to Implement an Emissions Cap-and-Trade Approach

The specific provisions of the Clean Air Act that are potentially applicable to control emissions of the pollutants discussed here can largely be categorized as provisions relating to either state programs for pollution control under Title I (e.g., sections 107, 108, 109, 110, 115, 126, and Part D of Title I), or national regulation of stationary sources through technology-based standards (e.g., sections 111 and 112). None of these provisions easily lends itself to establishing market-based national or regional emissions cap-and-trade programs.[4]

The Clean Air Act provisions relating to state programs do not authorize EPA to require states to control air pollution through economically efficient cap-and-trade programs and do not provide full authority for EPA itself to impose such programs. Under certain provisions in Title I, such as section 110, EPA may facilitate regional approaches to pollution control and encourage states to cooperate in a regional, cost- effective emissions cap-and-trade approach (see Notice of Proposed Rulemaking: Finding of Significant Contribution and Rulemaking for Certain States in the Ozone Transport Assessment Group Region for Purposes of Reducing Regional Transport of Ozone, 62 F.R. 60318 (Nov. 7, 1997)). EPA does not have authority under Title Ito require states to use such measures, however, because the courts have held that EPA cannot mandate specific emission control measures for states to use in meeting the general provisions for attaining ambient air quality standards. *See Commonwealth of Virginia v. EPA*, 108 F.3d 1397 (D.C. Cir. 1997). Under certain limited circumstances where states fail to carry out their responsibilities under Title I of the Clean Air Act, EPA has authority to take certain actions, which might include establishing a cap-and-trade program.[5] Yet EPA's ability to invoke these provisions for federal action depends on the actions or inactions of the states.

Technology-based standards under the Act directed to stationary sources have been interpreted by EPA not to allow compliance through intersource cap-and-trade approaches. The Clean Air Act provisions for national technology-based standards under sections 111 and 112 require EPA to promulgate regulations to control emissions of air pollutants from stationary sources. To maximize the opportunity for trading of emissions within a source. EPA has defined the term "stationary source" expansively, such that a large facility can be considered a "source." Yet EPA has never gone so far as to define as a source a group of facilities that are not geographically connected, and EPA has long held the view that trading across plant boundaries is impermissible under sections 111 and 112. See, e.g., National Emission Standards for Hazardous Air Pollutants for Source Categories; Organic Hazardous Air Pollutants from the Synthetic Organic Chemical Manufacturing Industry, 59 Fed. Reg. 19402 at 19425-26 (April 22, 1994).

III. Conclusion

EPA's regulatory authority under the Clean Air Act extends to air pollutants, which, as discussed above, are defined broadly under the Act and include SO2, NOx, CO2, and mercury emitted into the ambient air. EPA has in fact already regulated each of these substances under the Act, with the exception of CO2. While CO2 emissions are within the scope of EPA's authority to regulate, the Administrator or has made no determination to date to exercise that authority under the specific criteria provide under any provision of the Act.

With the exception of the SO2 provisions focused on acid rain, the authorities potentially available for controlling these pollutants from electric power generating sources do not easily lend themselves to establishing market-based national or regional cap-and-trade programs, which the Administration favors for addressing these kinds of pollution problems. Under certain limited circumstances, where states fail to carry out their responsibilities under Title I of the Act, EPA has authority to take certain actions, which might include establishing a cap-and-trade program. However, such authority depends on the actions or inactions of the states.

ATTACHMENT B

Testimony of Gary S. Guzy General Counsel U.S. Environmental Protection Agency before a Joint Hearing of the Subcommittee on National Economic Growth, Natural Resources and Regulatory Affairs of the Committee on Government Reform and the Subcommittee on Energy and Environment of the Committee on Science U.S. House of Representatives

October 6, 1999

Thank you, Chairman McIntosh, Chairman Calvert, and Members of the Subcommittees, for the invitation to appear here today. I am pleased to have this opportunity to explain the U.S. Environmental Protection Agency's (EPA) views as to the legal authority provided by the Clean Air Act (Act) to regulate emissions of carbon dioxide, or CO2.

Before I do, however, I would like to stress, as EPA repeatedly has stated in letters to Chairman McIntosh and other Members of Congress, that the Administration has no intention of implementing the Kyoto Protocol to the United Nations Framework Convention on Climate Change prior to its ratification with the advice and consent of the Senate.1 As I indicated in my letter of September 17, 1999 to Chairman McIntosh, there is a clear difference between actions that carry out authority under the Clean Air Act or other domestic law, and actions that would implement the Protocol. Thus, there is nothing inconsistent in assessing the extent of current authority under the Clean Air Act and maintaining our commitment not to implement the Protocol without ratification.

Some brief background information is helpful in understanding the context for this question of legal authority. In the course of generating electricity by burning fossil fuels, electric power plants emit into the air multiple substances that pose environmental concerns, several of which are already subject to some degree of regulation. Both industry and government share an interest in understanding how different pollution control strategies interact. These interactions are both physical (strategies for controlling emissions of one substance can affect emissions of others) and economic (strategies designed to address two or more substances together can cost substantially less than strategies for individual pollutants that are designed and implemented independently). EPA has worked with a broad array of stakeholders to evaluate multiple-pollutant control strategies for this industry in a series of forums, dating back to the Clean Air Power Initiative (CAPI) in the mid-1990s. While the CAPI process focused on SO2 and NOx, a broad range of participants, including representatives of power generators, the United Mine Workers, and environmentalists, expressed support for inclusion of CO2 emissions, along with SO2, NOx, and mercury, in subsequent analyses. One conclusion that emerged from these analytical efforts is that integrated strategies using market-based "cap-and-trade" approaches like the program currently in place to address acid rain would be the most flexible and lowest cost means to control multiple pollutants from these sources.

On March 11, 1998, during hearings on EPA's FY 1999 appropriations, Representative DeLay asked the Administrator whether she believed that EPA had authority to regulate emissions of pollutants of concern from electric utilities, including CO2. She replied that the Clean Air Act provides such authority, and agreed to Representative DeLay's request for a legal opinion on this point.

Therefore, my predecessor, Jonathan Z. Cannon, prepared a legal opinion for EPA Administrator Carol Browner on the question of EPA's legal authority to regulate several pollutants, including CO2 emitted by electric power generation sources. The legal opinion requested by Rep. DeLay was completed on April 10, 1998. It addressed the Clean Air Act authority to regulate emissions of four pollutants of concern from electric power generation: nitrogen oxides (NOx), sulfur dioxide (SO2), mercury, and CO2. Because today's hearing is focused exclusively on CO2, I will summarize the opinion's conclusions only as they relate to that substance.

The Clean Air Act includes a definition of the term "air pollutant," which is the touchstone of EPA's regulatory authority over emissions. Section 302(g) defines "air pollutant" as

> any air pollution agent or combination of such agents, including any physical, chemical, biological, [or] radioactive . . . substance or matter which is emitted into or otherwise enters the ambient air. Such term includes any precursors to the formation of any air pollutant, to the extent that the Administrator has identified such precursor or precursors for the particular purpose for which the Willi "air pollutant" is used.

Mr. Cannon noted that CO2 is a "physical [and] chemical substance which is emitted into . . . the ambient air," and thus is an "air pollutant" within the Clean Air Act's definition. Congress explicitly recognized emissions of CO2 from stationary sources, such as fossil fuel power plants, as an "air pollutant" in section 103(g) of the Act, which authorizes EPA to conduct a basic research and technology program to include, among other things, "[i]mprovements in nonregulatory strategies and technologies for preventing or reducing multiple *air pollutants, including* sulfur oxides, nitrogen oxides, heavy metals, PM-10 (particulate matter), carbon monoxide, *and carbon dioxide,* from stationary sources, including fossil fuel power plants." (Emphasis added.)

The opinion explains further that the status of CO2 as an "air pollutant" is not changed by the fact that CO2 is a constituent of the natural atmosphere. In other words, a substance can be an "air pollutant" under the Clean Air Act's definition even if it has natural sources in addition to its man-made sources. EPA regulates a number of naturally-occurring substances as air pollutants because human activities have increased the quantities present in the air to levels that are harmful to public health, welfare, or the environment. For example, SO_2 is emitted from geothermal sources; volatile organic compounds (VOCs), which are precursors to harmful ground-level ozone, are emitted by vegetation. Some substances regulated under the Act as hazardous air pollutants are actually necessary in trace quantities for human life, but are toxic at higher levels or through other routes of exposure. Manganese and selenium are two examples of such pollutants. Similarly, in the water context, phosphorus is regulated as a pollutant because although it is a critical nutrient for plants, in excessive quantities it kills aquatic life in lakes and other water bodies.

While CO_2, as an "air pollutant," is within the scope of the regulatory authority provided by the Clean Air Act, this by itself does not lead to regulation. The Clean Air Act includes a number of regulatory provisions that may potentially be applied to an air pollutant. But before EPA can actually issue regulations governing a pollutant, the Administrator must first make a formal finding that the pollutant in question meets specific criteria laid out in the Act as prerequisites for EPA regulation under its various provisions. Many of these specific Clean Air Act provisions for EPA action share a common feature in that the exercise of EPA's authority to regulate air pollutants is linked to a determination by the Administrator regarding the air pollutant's actual or potential harmful effects on public health, welfare or the environment. For example, EPA has authority under section 109 of the Act to establish National Ambient Air Quality Standards for any air pollutant for which the Administrator has established air quality criteria under section 108. Under section 108, the Administrator must first find that the air pollutant in question meets several criteria, including that:

it causes or contributes to "air pollution which may reasonably be anticipated to endanger public health or welfare;" and

its presence in the ambient air "results from numerous or diverse mobile or stationary sources . . . ."

Section 302(h), a provision dating back to the 1970 version of the Clean Air Act, defines "welfare" and states:

> all language referring to effects on welfare includes, but is not limited to, effects on soils, water, crops, vegetation, man-made materials, animals, wildlife, weather, visibility, and climate, damage to and deterioration of property, and hazards to transportation, as well as effects on economic values

and on personal comfort and well-being, whether caused by transformation, conversion, or combination with other air pollutants.

Thus, since 1970, the Clean Act has included effects on "climate" as a factor to be considered in the Administrator's decision, as to whether to list an air pollutant under section 108.

Analogous threshold findings are required before the Administrator may establish new source performance standards for a pollutant under section 111, list and regulate the pollutant as a hazardous air pollutant under section 112, or regulate its emission from motor vehicles under Title II of the Act.

Given the clarity of the statutory provisions defining "air pollutant" and providing authority to regulate air pollutants, there is no statutory ambiguity that could be clarified by referring to the legislative history. Nevertheless, I would note that Congress' decision in the 1990 Amendments not to adopt additional provisions directing EPA to regulate greenhouse gases by no means suggests that Congress intended to limit *pre-existing* authority to address any air pollutant that the Administrator determines meets the statutory criteria for regulation under a specific provision of the Act.

I would like today to reiterate one of the central conclusions of the Cannon memorandum, which stated: "While CO2, as an air pollutant, is within EPA's scope of authority to regulate, the Administrator has not yet deteiiuined that CO2 meets the criteria for regulation under one or more provisions of the Act." That statement remains true today. EPA has not made any of the Act's threshold findings that would lead to regulation of CO2 emissions from electric utilities or, indeed, from any source. The opinion of my predecessor simply clarifies — and I endorse this opinion — that CO2 is in the class of compounds that could be subject to several of the Clean Air Act's regulatory approaches. Thus, I would suggest that many of the concerns raised about the statutory authority to address CO2 relate more to factual and scientific, rather than legal, questions regarding whether and how the criteria for regulation under the Clean Air Act could be satisfied.

I also want to note, however, EPA has strongly promoted voluntary partnerships to reduce emissions of greenhouse gases through the EnergyStar and Green Lights programs and other non-regulatory programs that Congress has consistently supported. These successful programs already have over 7,000 voluntary partners who are taking steps to reduce greenhouse gas emissions, reduce energy costs and help address local air pollution problems. These programs also help the United States meet its obligations under the United Nations Framework Convention on Climate Change, which was ratified in 1992. I would also note, as EPA has indicated in past correspondence with Chairman McIntosh and others, in the course of carrying out the mandates of the Clean Air Act, EPA has in a few instances directly limited use or emissions of certain greenhouse gases other than CO2. For example, EPA has limited the use of certain substitutes for ozone-depleting substances under Title VI of the Act, where those substitutes have very high global warming potentials. I wish to stress once more, however, that while EPA will pursue efforts to address the threat of global warming through the voluntary programs authorized and funded by Congress and will carry out the mandates of the Clean Air Act, this Administration has no intention of implementing the Kyoto Protocol prior to its ratification on the advice and consent of the Senate.

This concludes my prepared statement. I would be happy to answer any questions that you may have.

End Notes

[1] See also section 103(g) of the Act (authorizes EPA to conduct a basic research and technology program to develop and demonstrate nonregulatory strategies and technologies for air pollution prevention, which shall include among the program elements "[flm.provements in nonregulatory strategies and technologies for preventing or reducing multiple air pollutants, including sulfur oxides, nitrogen oxides, heavy metals, PM-10 (particulate matter), carbon monoxide, and carbon dioxide, from stationary sources, including fossil fuel power plants.")

[2] See. e g., section 108 (directs Administrator to list and issue air quality criteria for each air pollutant that causes or contributes to air pollution that may reasonably be anticipated to endanger public health or welfare and that is present in the ambient air due to emissions from numerous or diverse mobile or stationary sources); section 109 (directs Administrator to promulgate national primary and secondary ambient air quality standards for each air pollutant for which there are air quality criteria, to be set at levels requisite to protect the public health with an adequate margin of safety (primary standards) and to protect welfare (secondary standards)), Section 110 (requires States to submit state implementation plans (S1Ps) to meet standards); Section 111 (b) (requires Administrator to list, and set federal performance standards for new sources in, categories of stationary sources that cause or contribute significantly to air pollution that may reasonably be anticipated to endanger public health or welfare); section 111(d) (states must establish performance standards for existing sources for any air pollutant (except criteria pollutants or hazardous air pollutants) that would be subject to a performance standard if the sources were a new source), section. 112(b) (lists 188 hazardous air pollutants and authorizes Administrator to add pollutants to the list that may present a threat of adverse human health effect or adverse environmental effects); section 112(d) (requires Administrator to set emissions standards for each category or subcategory of major and area sources that the Administrator has listed pursuant to section 119(c)); section 112(n)(1)(A) (requires Administrator to study and report to Congress on the public health hazards reasonably anticipated from emissions of limited hazardous air pollutants from electric utility steam generating units, and requires regulation if appropriate and necessary); section 115 (Administrator may require state action to control

certain air pollution if, on the basis of certain reports, she has reason to believe that any air pollutant emitted in the United States causes or contributes to air pollution that may be reasonably anticipated to endanger public health or welfare in a foreign country that has given the United States reciprocal rights regarding air pollution control) Title IV (establishes cap-and-trade system for control of SO2 from electric power generation facilities and provides for certain controls on NOx).

[3] The language in Section 302(h) listing specific potential effects on welfare, including the references to weather and climate, dates back to the 1970 version of the Clean Air Act.

[4] Title 1V of the Act provides explicit authority for a cap and trade program for SO2 emissions from electric power generating sources.

[5] For example, section 110(c) requires EPA to promulgate a Federal implementation plan where EPA finds that a state has failed to make a required submission of a SIP or that the SIP or SIP revision does not satisfy certain minimum criteria, or EPA disapproves the SIP submission in whole or in part in addition, section 126 provides that a State or political subdivision may petition the Administrator for certain findings regarding emissions from certain stationary sources in another state. If the Administrator grants the petition, she may establish control requirements applicable to sources that were the subject of the petition.

In: Climate Change Litigation and Law
Editors: Jean-François Masson ISBN: 978-1-60876-089-3

Chapter 5

HEARING ON IMPLICATIONS OF SUPREME COURT'S EPA DECISION- DONIGER TESTIMONY

David Doniger
Committee on Environment and Public Works
United States Senate

Madame Chairman, members of the Committee, thank you for the opportunity to testify today on *Massachusetts v. EPA,* the Supreme Court's decision upholding the Environmental Protection Agency's authority to regulate global warming pollution under the Clean Air Act. I am policy director and senior attorney for the Natural Resources Defense Council's Climate Center. I represent NRDC and its 1.2 million members and supporters in the *Massachusetts* case and in related global warming litigation. I work closely with the broad coalition of states, cities, and environmental organizations engaged in these cases. In the 1990s, I served as director of climate change policy in the EPA air office, under Carol Browner.

We began this case during the coldest part of the Little Ice Age in global warming policy in Washington. The President had broken his campaign pledge to control carbon dioxide. The Congress was inactive. The states were not yet moving. Yet the science was growing ever clearer on the dangers of global warming, and the nation's Clean Air Act already empowered the government to react to that science. When the Bush Administration tried to nail this door permanently closed, our coalition of states, cities, and environmental organizations took the last step available, appealing to the independent third branch to uphold our nation's laws.

The Supreme Court's April 2nd decision in *Massachusetts v. EPA* repudiates the Bush Administration's legal strategy for doing nothing on global warming. The nation's highest court set the White House straight: Carbon dioxide is an air pollutant. EPA has – and has always had – the power and responsibility to start cutting the pollution that is wreaking havoc with our climate. We need EPA to act *now*.

The Court's decision has four immediate game-changing consequences:

First, Administrator Johnson now must decide afresh whether to set greenhouse gas emission standards for new motor vehicles under Section 202 of the Clean Air Act. The Court clearly stated that this decision must be based on the science, and the science only:

> Under the clear terms of the Clean Air Act, EPA can avoid taking further action only if it determines that greenhouse gases do not contribute to climate change or if it provides some reasonable explanation as to why it cannot or will not exercise its discretion to determine whether they do.[1]

The Court rejected all of the Administration's "laundry list of reasons not to regulate" – preferences for voluntary action, concerns about piecemeal regulation, claimed interference with foreign policy. No, the Court said, the decision must be made on the science only: "To the extent that this constrains agency discretion to pursue other priorities of the Administrator or the President, this is the congressional design."[2]

The Court was especially clear that "while the President has broad authority in foreign affairs, that authority does not extend to the refusal to execute domestic laws."[3] The Court also observed that while economic considerations figure into the level and timing of standards under the Clean Air Act, they are not relevant to determining the *need* for such standards.[4]

The Court found that "EPA has not identified any congressional action that conflicts in any way with the regulation of greenhouse gases from new motor vehicles."[5] Specifically, the Court found no conflict with the Energy Policy and Conservation Act, under which the Corporate Average Fuel Economy (CAFE) standards are set: "[T]hat DOT sets mileage standards in no way licenses EPA to shirk its environmental responsibilities. EPA has been charged with protecting the public's "health" and "welfare," . . . a statutory obligation wholly independent of DOT's mandate to promote energy efficiency."[6]

Given the Administration's embrace of the Intergovernmental Panel on Climate Change (IPCC) – more than 90 percent certainty that anthropogenic emissions are causing global warming – it is difficult to imagine how Administrator Johnson could not now conclude that vehicular emissions of these pollutants are contributing to climate change. He must act, and now.

Since the Supreme Court's decision, however, the Administration has made statements that give reason for concern about their intentions to comply with the Court's decision. On April 3rd, while acknowledging that the Court's decision is "the new law of the land," President Bush himself went right back to the well of extraneous considerations that the Court one day before had declared illegal: research, voluntary action, waiting for other countries to act.[7] Administrator Johnson sounded the same notes at a press conference on April 10th . And Council on Environmental Quality Chairman James Connaughton declared the Court's decision "somewhat moot" and "inconsequential" because "the President is already committed to regulatory action" – by which he meant that the Administration had *asked Congress for new laws* on fuel economy and alternative fuels.[8]

This Committee will no doubt hear about a long list of voluntary programs and initiatives. Some of EPA's programs, such as EnergySTAR labeling, have brought about real changes in the energy-consuming products we purchase. Other programs, such as the Asia-Pacific Partnership (on which I have testified before), are utterly ineffective efforts just to look busy.

Altogether, these voluntary efforts have failed to stop the steady growth in U.S. emissions, which has continued during the Bush years at the same rate as in the prior decade – about 14 percent per decade. The Administration cloaks its statistics in the deceptive metric of "emissions intensity." Celebrating improvements in emissions intensity is like a dieter's claiming victory when he succeeds only in slowing his weight gain.

This Committee has a special role and responsibility to hold Administrator Johnson's feet to the fire. Demand that Mr. Johnson give you a specific schedule for determining that vehicles' heat-trapping emissions are in fact contributing to global warming. Do not accept procedural dodges and delays. There is no more legal basis, and no more time, for these lame excuses.

Second, and equally important, the *Massachusetts* decision removes the major obstacle to *state* initiatives, led by California, to cut global warming pollution from vehicles. California and 11 other states – Connecticut, Maine, Maryland, Massachusetts, New Jersey, New York, Oregon, Pennsylvania,

Rhode Island, Vermont, and Washington – have adopted clean car standards to cut heat-trapping emissions by 30 percent by model year 2016. Arizona and New Mexico – and perhaps others – will soon join. Together, these states account for more than a third of the U.S. vehicle market. The Clean Air Act allows California to set its own air pollutant standards, provided only that it gets a routine waiver from EPA. California asked for the waiver nearly 16 months ago, but EPA has been dragging its feet.

I welcome Administrator Johnson's recent commitment to Governor Schwarzenegger that he will now allow the waiver process to start, and that EPA will soon publish a notice scheduling the required hearing. But the Administrator has declined to give California any schedule for making the waiver decision itself. The standards apply starting in the 2009 model year, which is fast approaching. This Committee should demand a clear and near-term deadline from Administrator Johnson for his decision. It is time for the Administration to stop stalling and get out of California's way.

Third, the Supreme Court's decision has implications for other pending global warming litigation. At the top of the list is a parallel case on power plants. A coalition of states and environmental organizations has challenged EPA's refusal to add a CO_2 emission standard to the new source performance standards and emission guidelines for new and existing power plants under Section 111 of the Clean Air Act in a case called *New York v. EPA*.[9] EPA's sole reason for refusing to regulate was the claim that it had no legal authority to control CO_2 – the very issue now settled by *Massachusetts*. The D.C. Circuit stayed that case pending the Supreme Court's decision, and now we intend to seek an immediate reversal of the EPA position.

So Administrator Johnson now will also have to decide whether CO_2 emissions from power plants contribute to global warming. Again, based on the clear scientific evidence, we cannot see how he could reach any other conclusion. As with vehicles, Administrator Johnson must act on power plants, and now. We hope this Committee will press him for action here too.

The *Massachusetts* decision will very helpfully affect other cases also. It knocks the legs out from under cases brought by the auto industry in California, Vermont, and Rhode Island, alleging that those states lack Clean Air Act authority to set clean car standards, and alleging conflict with the CAFE standards. *Massachusetts* also strengthens the states' position in *Connecticut v. American Electric Power*, a case pending in the Second Circuit Court of Appeals. In that case, eight states, New York City, and two land conservation trusts allege that the five electric power companies with the highest CO_2 emissions are creating a public nuisance. Their theory stems directly from *Georgia v. Tennessee Copper,*[10] the case relied on by the Supreme Court in *Massachusetts* to buttress states' standing and states' right to go to federal court to abate pollution outside their borders.

Fourth, and most important, the Supreme Court's decision has added new momentum to the legislative process. Even before April 2nd, the legislative kettle was nearing a boil. Since Hurricane Katrina, and since the November elections, public sentiment has shifted dramatically on global warming. Congress's new leaders and committee chairs have expressed the strong commitment to pass comprehensive global warming legislation. Many forward-looking business leaders have come forward to embrace the desirability – or at least the inevitability – of new legislation. Perhaps motivated by the prospect that this Administrator – or the next one – will use his Clean Air Act powers, even more industry leaders are coming to the table now to help hammer out new global warming legislation. As they say, "If you're not at the table, you're on the menu."

NRDC supports placing every ounce of pressure you can on the Administration to faithfully execute the existing law of the land. The actions already within EPA's power would take a big bite out of global warming. At the same time, we also support enactment of new economy-wide legislation to comprehensively address global warming.

In NRDC's view, solving global warming requires three things:

- A mandatory declining cap on national emissions that starts cutting emissions now and reduces them by 80% by 2050.
- Performance standards – for vehicles, fuels, and power plants, as well as buildings, appliances, and other equipment – to quickly deploy today's emission-cutting technology and promote rapid development of tomorrow's.
- Incentives – drawn mainly from the value of emissions allowances – to promote new technology, to protect consumers (especially low-income citizens), workers, and communities, and to help manage adaptation to climate impacts that we cannot avoid.

There is still time – though only a little time – to avoid the worst effects of global warming. If the United States and other industrial countries commit to action on this scale, and if key developing countries also reduce their emissions growth and follow suit with similar reductions later in the century, then we can still keep greenhouse gas concentrations from exceeding 450 parts per million (CO_2-equivalent) and maintain at least a 50-50 chance of avoiding warming of more than another 2 degrees Fahrenheit. Exceeding this level, more and more scientists tell us, is extremely dangerous.

NRDC looks forward to working with this Committee and with all stakeholders to pass this legislation in this Congress.

End Notes

[1] 127 S.Ct. 1438, 1462 (2007).
[2] *Id.*
[3] *Id.* at 1462-63.
[4] *Id.* at 1461.
[5] *Id.* at 1461.
[6] *Id.* at 1462.
[7] http://www.whitehouse.gov/news/releases/2007/04/20070403.html.
[8] Jeff Bernard (AP), "Rulings Go Against Bush Administration," Casper Star-Tribune (Apr. 18, 2007) (attached).
[9] No. 06-1322 (D.C. Cir.)
[10] 206 U.S. 230 (1907).

In: Climate Change Litigation and Law
Editors: Jean-François Masson

ISBN: 978-1-60876-089-3

Chapter 6

HEARING ON IMPLICATIONS OF SUPREME COURT'S EPA DECISION-GLASER TESTIMONY

Peter Glaser
Senate Committee on Environment and Public Works

My name is Peter Glaser. I am a partner in the Washington, D.C., office of Troutman Sanders LLP. I received a B.A. from Middlebury College in 1975 and a J.D. from the George Washington University National Law Center in 1980. I practice in the areas of environmental and energy law. I represented the Washington Legal Foundation in filing an amicus brief before the Supreme Court in the *Massachusetts v. EPA* litigation.

Let me begin by stating that I am not here before the committee representing or advocating the position of any particular company or industry. I am not receiving remuneration from anyone for my testimony, and the views expressed in my testimony are my own and not necessarily those of any company or group that I currently represent or have represented.

In addition, I am not here to recommend any particular course of action by this Committee or Congress. I have been asked to offer my views as a practicing attorney of issues pertaining to the potential regulation of greenhouse gases (GHGs) for global warming purposes by the U.S. Environmental Protection Agency (EPA) under the Clean Air Act (CAA).

Under the Court's decision in *Massachusetts v. EPA*, EPA will be required to decide whether GHGs emitted by new motor vehicles may reasonably be anticipated to endanger public health or welfare. The Court did not require EPA to make an endangerment finding. It remanded the case to EPA for further consideration of the endangerment issue. Therefore, based on its analysis of the science, EPA's options are to make an endangerment finding, make a non-endangerment finding, or decide that the science is insufficiently certain to decide either way.

Although the *Massachusetts* case concerned potential regulation of GHG emissions from new motor vehicles under Title II of the CAA, there is no doubt that the Court's ruling that GHGs are "air pollutants" under the statute may have implications for other CAA regulatory programs. Indeed, EPA was asked to set New Source Performance Standards for carbon dioxide (CO2) from fossil-fuel-fired electric generating units under Section 111 of the CAA. EPA declined that request last year, stating it had no authority to regulate GHGs for global warming purposes, and the matter was appealed to and is pending in the U.S. Court of Appeals for the D.C. Circuit. It can now be expected that the case will be remanded to EPA for further action in light of the *Massachusetts v. EPA* decision.

Thus, it is likely that EPA will have two formal cases before it in the near term in which it will be examining potential GHG regulation. One of the cases will address motor vehicles under Title II, and the other will address electric generating units under Title I. In addition, it is possible that EPA on its own motion or in response to further petitions may consider potential GHG regulation for other sources.

However, the character of any such regulation remains uncertain. Although the Court's decision clearly provides for EPA regulation under Title II if an endangerment finding is made, the decision does not say anything at all about what that regulation should be or when it should become effective. Those matters are left to EPA judgment confined by the specific CAA provisions under which EPA would invoke regulation. One of the arguments made by EPA and supporting parties in the *Massachusetts* litigation was that the CAA was not designed to address an issue such as global climate change. While the Court ruled that GHGs meet the CAA's definition of "air pollutant," the fact remains that GHG regulation under the CAA is likely to be an uncomfortable fit.

The most obvious example is the National Ambient Air Quality Standards (NAAQS) program, the program the Courts have termed the "cornerstone" of Title I of the CAA. One of the prerequisites for the establishment of air quality criteria and NAAQS in Sections 108 and 109 of the CAA is similar to the regulatory trigger language the court construed in *Massachusetts*. Yet it is hard to imagine how NAAQS regulation would work for a GHG. The establishment of a NAAQS triggers a process whereby attainment and nonattainment areas are designated, states are required to submit implementation plans to attain or maintain the NAAQS, and severe sanctions are mandated for non-compliance. Yet, given the nature of globally-circulating GHGs, where a ton of GHG emitted in, for instance, Maryland has the same impact on GHG concentrations over Maryland as a ton emitted in China, there is nothing Maryland could do about attaining or maintaining a GHG NAAQS. Maryland could literally cease emitting any GHGs tomorrow and it would have no discernable impact on GHG concentrations over the state.

Similarly, GHG emissions are not a pollutant transport issue, such as ozone, where groups of states can combine to reduce emissions for the purpose of regional attainment. Given the nature of the issue, not even the most draconian multi-state emission reductions could ensure attainment or maintenance of a GHG NAAQS.

I do not conclude that, if EPA makes an endangerment finding for motor vehicles under Title II, it has authority to establish a GHG NAAQS since the trigger language in Section 108 is not identical to the Section 202 trigger language construed in *Massachusetts*. Nevertheless, given the similarities, it is not a stretch to imagine a petition alleging that EPA not only has authority to establish a NAAQS, it must establish a NAAQSs. That issue would be a difficult one for the agency and the courts to resolve.

Perhaps a more likely initial battleground for EPA CAA regulation, assuming an endangerment finding is made, is the NSPS program under Section 111. Yet this program too is likely to create regulatory difficulties. A first issue might be whether Section 111 authorizes EPA to create a market-based cap and trade program, or whether EPA's authority is limited to imposing more inefficient command-and-control technology requirements on individual sources. In the Clean Air Mercury Rule, now being litigated before the U.S. Court of Appeals for the D.C. Circuit, EPA interpreted Section 111 as allowing it to implement a cap and trade program to control mercury emissions from coal- and oil-fired utility units. However, a group of environmental parties has filed a brief challenging EPA's authority to utilize a cap and trade program under Section 111, claiming that a cap and trade program does not meet the definition of a "standard of performance" under that section. Thus, the ability to utilize cap and trade under Section 111 is, at least for the moment, uncertain.

Section 111 creates additional regulatory difficulties for controlling GHG emissions. A "standard of performance" is defined under Section 111 (a)(1) as "a standard for emissions of air pollutants which reflects the degree of emission limitation achievable through the application of the best system of emission reduction which (taking into account the cost of achieving such reduction and any nonair quality health and environmental impact and energy requirements) the Administrator determines has been adequately demonstrated." This standard has come to be known as "best demonstrated technology" or BDT. As can be seen, under BDT, both the availability of technology and the cost of technology are factors the Administrator must consider in setting a standard of performance. It is true that the standard can be set to be "technology forcing." On the other hand, the standard cannot be based on results achieved short-term at a small-scale "pilot" plant. EPA must show that the standard is "achievable" in the real world, that is, it "must be 'adequately demonstrated' that there will be 'available technology.'" *Portland Cement Ass'n v. Ruckelshaus*, 486 F.2d 375, 391 (D.C. Cir. 1973), *cert. denied*, 417 U.S. 921 (1974), quoting the statutory text. It is EPA's burden to make this demonstration; it cannot be passed off to industry. *National Lime Ass 'n v. EPA*, 627 F.2d 416, 432 (D.C. Cir. 1980).

These standards will be difficult to apply to the nation's coal-fired electric generation fleet. While I do not offer myself up as an expert on carbon control technologies, cost-effective technologies do not appear to exist *today* for controlling carbon emissions from coal-based electric generating plants on a large-scale basis. Certainly many promising technologies are in development, and both the Department of Energy and the Electric Power Research Institute expect these technologies will become cost-effective at some point after 2020. But for purposes of developing standards of performance for coal-based generation today, new source performance standards are likely to prove controversial. For instance, carbon-scrubbing at a pulverized coal plant may consume a very large percentage of that unit's total electric power. This is likely to be problematic given the requirement in determining standards of performance for considering the energy requirements of the control technology.

Controlling emissions from coal-based generation through the NSPS program is also likely to prove difficult because of the need not only to capture carbon dioxide but to store it safely indefinitely. Again, the results of initial testing are promising and, in the not too distant future, sufficient testing is expected to be accomplished to demonstrate the ability to store large quantities of carbon dioxide underground over the long-term. In the meantime, however, given the lack of large-scale storage data, and the very difficult liability issues presented by underground storage, an attempt to establish a standard of performance for carbon capture and storage may be difficult to justify.

Other possibilities for application of the NSPS program to control carbon emissions from the electric power sector might be requirements for the use of IGCC technology or even fuel switching to natural gas. Even under Section 111(h), there are significant legal issues as to whether such requirements would be valid. Section 111(h) provides that, if EPA determines that it not feasible to prescribe a standard of performance, EPA may prescribe "a design, equipment, work practice, or operational standard, or combination thereof, which reflects the best technological system of continuous emission reductions which (taking into consideration the cost of achieving such emission reduction, and any non-air quality health and environmental impact and energy requirements) the Administrator determines has been adequately demonstrated." Use of this section to, for instance, set a standard for carbon emissions from a coal plant that would require switching to natural gas would be unprecedented. EPA has regulated sulfur dioxide and nitrogen oxide emissions from coal plants for many years, but has never determined that gas plant emissions should set the standard for emissions from coal such that coal would have to be replaced by gas. It is hard to imagine EPA attempting to utilize Section 111 to, in essence, order that coal plants convert to gas technology. The economic impacts of such decisions could be staggering.

The fact that new source performance standards must be technology- and cost-based creates further difficulties in utilizing Section 111 to implement a cap and trade program for GHG emissions, even assuming a cap and trade program represents a valid standard of performance. In CAMR, for instance, EPA was constrained in choosing the two-phase mercury caps by the Section 111 requirement that a standard of performance be achievable. EPA's methodology for calculating the cap thus involved essentially determining the mercury emission reductions achievable at individual units, summing those reductions up nationwide, and setting the cap on that basis. A GHG cap would have to be set on the same basis, that is, based on a determination of what is achievable nationwide based on technology and cost considerations. EPA could not simply choose a cap based solely on its views of desirable emission reductions.

Finally, the U.S. Court of Appeals for the D.C. Circuit, has stated that the Best Demonstrated Technology standard is a very broad standard indeed. According to the Court, "[t]he language of section 111 . . . gives EPA authority . . . to weigh cost, energy, and environmental impacts in the broadest sense at the national and regional levels and over time as opposed to simply at the plant level in the immediate present." *Sierra Club v. Costle*, 657 F.2d 298, 330 (D.C. Cir. 1981). The Court stated that "'section 111 of the Clean Air Act, properly construed, requires the functional equivalent of a NEPA impact statement.'" *Id.* at 331, quoting *Portland Cement*, 486 F.2d at 384.

Moreover, in 1980, in a case involving the limestone industry, the Court noted the "rigorous standard of review under section 111" applied by reviewing courts. *National Lime*, 627 F.2d at 429. The Court stated that the "sheer massiveness of impact of the urgent regulations," considered in that and other cases had "prompted the courts to require the agencies to develop a more complete record and a more clearly articulated review for arbitrariness and caprice" than had been applied in previous cases. *Id.* at 451 n.126.

If massiveness of regulatory impact was a concern in a limestone industry case, that concern would be magnified many times in promulgating GHG standards of performance. A plethora of issues would be relevant in setting GHG standards, with EPA weighing the cost, energy and, and environmental impacts of GHG regulation "in the broadest sense at the national and regional levels and over time" as if it were preparing an Environmental Impact Statement. A large number of parties would be interested given the overweening importance of the issues.

Thus, an EPA rulemaking to establish NSPS for utility units would be highly complex, controversial and time-consuming. Quick results, to say the least, cannot reasonably be expected.

In conclusion, back when the issue that ultimately led to the *Massachusetts* decision first began, then EPA General Counsel Jonathan Z. Cannon wrote an April 10, 1998 memorandum to then Administrator Carol M. Browner examining potential regulation of GHGs under various provisions of the CAA. He concluded that "[n]one of these provisions easily lends itself to market-based national or regional emissions cap-and-trade programs." It is also true that attempting to utilize Section 111 to control the nation's GHG emissions, either through command or control or cap and trade, would be complicated and controversial. In the aftermath of the *Massachusetts* decision, EPA may undertake proceedings to determine whether a sound basis exists to make an "endangerment" finding and, if so, to then determine what kind of regulations it may intend to propose under which specific CAA program. But the ability of EPA to utilize the CAA to create an ambitious regulatory regime is likely to prove very difficult indeed.

In: Climate Change Litigation and Law
Editors: Jean-François Masson

ISBN: 978-1-60876-089-3

Chapter 7

HEARING ON IMPLICATIONS OF SUPREME COURT'S EPA DECISION- KLEE TESTIMONY

Ann R. Klee
Senate Committee on Environment and Public Works

The Implications of the Supreme Court's Decision Regarding EPA's Authorities with Respect to Greenhouse Gas Emissions under the Clean Air Act April 24, 2007

Madam Chairman, Members of the Committee, I am pleased to be here today to testify on the issue of EPA's authority to regulate greenhouse gas emissions under the Clean Air Act in the wake of the Supreme Court's decision in *Massachusetts v. EPA*, No. 05-1120, 549 U.S. ____ (2007). Before I begin, however, I would like to make clear that my testimony today reflects my personal views and analysis of the law based upon my experience as General Counsel of the U.S. Environmental Protection Agency (from 2004 until August 2006), as Chief Counsel of this Committee (from 1997 until 2000), and most recently as a lawyer in private practice.

OVERVIEW

On April 2, 2007, the United States Supreme Court issued its landmark vacating the Environmental Protection Agency's (EPA's) denial of a petition to regulate greenhouse gas emissions from new motor vehicles. The Court's majority found that greenhouse gas emissions are "air pollutants" under the Clean Air Act and, therefore, potentially subject to regulation if, *in the judgment of the Administrator*, they "cause or contribute to air pollution which may reasonably be anticipated to endanger public health or welfare." The Court did not reach the issue of whether greenhouse gases from new motor vehicles, in fact, endanger public health or welfare, but it potentially significantly constrained EPA's discretion with respect to that determination.

Undoubtedly, the decision has changed the regulatory landscape. The determination that greenhouse gases are air pollutants will likely lead EPA to regulate greenhouse gas emissions, and carbon dioxide (CO_2) in particular, from new motor vehicles. It also likely will lead to regulation of stationary sources of greenhouse gases since the Clean Air Act's stationary source provisions are also triggered by an "endangerment" finding. In this respect, the decision is a significant one – an endangerment finding under one program will make it very difficult for EPA not to regulate under other programs.

The decision will not, however, have any meaningful impact in terms of addressing global climate change. Forcing the square peg of greenhouse gas emissions through the round holes of EPA's existing regulatory tools – tailpipe standards, national ambient air quality standards, new source performance standards, *etc.* – may have the effect of reducing U.S. emissions over time, but it will do nothing to reduce atmospheric concentrations of greenhouse gases, which is the true measure of effectiveness of regulation for climate change purposes. Unless our trading partners, China and India in particular, are also part of the effort to reduce global emissions of greenhouse gases, piece-meal regulation in the United States will not only achieve little; it may, in fact, have the unintended effect of leading to increased emissions by encouraging the relocation of U.S. businesses to countries not subject to greenhouse gas regulation.

THE *MASSACHUSETTS V. EPA* DECISION

The *Massachusetts* case involved a challenge to EPA's denial of a petition to regulate greenhouse gas emissions from new motor vehicles under section 202 of the Clean Air Act. EPA denied the petition on the grounds that it lacked the authority under the Act to regulate emissions for climate change purposes and, in the alternative, that even if it had the authority to set greenhouse gas standards, it would not be "effective or appropriate" to do so at this time. On appeal to the Supreme Court, petitioners raised two central questions: (1) whether EPA has the statutory authority to regulate greenhouse gas emissions from new motor vehicles under section 202 of the Clean Air Act; and (2) if the Agency does have the authority, whether its stated reasons for declining to regulate greenhouse gas emissions from new motor vehicles was consistent with the statute.[1]

Writing for the Majority in a 5-4 decision, Justice Stevens answered the first question in the affirmative, concluding that the Clean Air Act's language is unambiguous and that carbon dioxide is an "air pollutant" within the meaning of the Act and, therefore, potentially subject to regulation. Justice Stevens went on to reject the basis upon which EPA had decide not to regulate greenhouse gas emissions at this time. Justice Scalia filed a dissenting opinion on the merits on behalf of himself, Chief Justice Roberts, and Justices Alito and Thomas. The dissenting opinion reached the opposite conclusion with respect to both questions.

The term "air pollutant" is defined in the statute as "any air pollution agent or combination of such agents, including any physical, chemical, . . . substance or matter which is emitted into or otherwise enters the ambient air." Focusing solely on the language following the word "including," Justice Stevens adopts the view that carbon dioxide is a chemical or physical substance emitted into the air and must therefore be an air pollutant.[2] His opinion does not address whether carbon dioxide meets the first element of the definition, namely whether it is first an "air pollution agent." As EPA argued in its brief, and as Justice Scalia noted in his dissenting opinion, the fact that the statutory definition uses the words "any" and "including" does not end the analysis. As he points out, "in order to be an 'air pollutant' under the Act's definition, the 'substance or matter [being] emitted into the . . . ambient air' must also meet the first half of the definition – namely it must be an "air pollution agent or combination of such agents." The phrase following the term "including" can be illustrative of the kind of substances that might also be air pollution agents, but does not necessarily substitute for the first element of the definition. EPA provided the following example, quoted by Justice Scalia, in support of this point: "The phrase 'any American automobile, including any truck or minivan,' would not naturally be construed to encompass a foreign-manufactured [truck or] minivan." Scalia Dissent at 9.

Having concluded that greenhouse gas emissions are "air pollutants" within the meaning of the statute, Justice Stevens has "little trouble concluding" that EPA is "authorize[ed] to regulate greenhouse gas emissions from new motor vehicles in the event that it forms a "judgment" that such emissions contribute to climate change." Slip op. at 25. Section 202(a)(1) of the Act provides that EPA "shall by regulation prescribe . . . standards applicable to the emission of any air pollutant from any class or classes of new motor vehicles or new motor vehicle engines, which in [the Administrator's] judgment cause, or contribute to, air pollution which may reasonably be anticipated to endanger public health or welfare." To date, EPA has never made an endangerment finding with respect to carbon dioxide.

Finally, the Court rejected EPA's alternative basis for its decision not regulate greenhouse gas emissions from new motor vehicles at this time. EPA had argued that even if the Clean Air Act did authorize the Agency to regulate greenhouse gas emissions from new motor vehicles, that it appropriately exercised its discretion not to make an endangerment finding and regulate those emissions at this time. The Agency based its decision on, among other things, the continuing scientific uncertainties that were summarized in a 2001 National Academy of Sciences Report, as well as legitimate policy considerations, including the President's comprehensive approach to addressing climate change through investment in technology and voluntary actions. As EPA noted, "establishing [greenhouse gas] emissions standard for U.S. motor vehicles at this time would . . . result in an inefficient, piecemeal approach to addressing the climate change issue. . . . .A sensible regulatory scheme would require that all significant sources and sinks of [greenhouse gas] emissions be considered in deciding how best to achieve any needed emissions reductions." 68 Fed. Reg. 52,929-931.

The Court, however, concluded that EPA"s exercise of its "judgment" in this case was based upon "reasoning divorced from the statutory text" and therefore invalid. Slip op. at 30. Even though the statute is silent with respect to how the Agency shall exercise its "judgment" in the context of an endangerment finding, and even though the term "endanger" is not defined in the statute, the Court substantially constrained the Agency's ability to exercise its judgment, at least with respect to a determination under section 202 of the

Act. In effect, the Court held that "EPA can avoid taking further action only if it determines that greenhouse gas emissions do not contribute to climate change or if it provides some reasonable explanation as to why it cannot or will not exercise its discretion to determine whether they do." Slip op. at 30. With respect to the latter, the Court suggests that the only basis for not exercising its discretion would be if "the scientific uncertainty is so profound that it precludes EPA from making a reasoned judgment as to whether greenhouse gases contribute to global warming." Slip op. at 31.

Significantly, the Court did not reach the question of whether EPA must actually make an endangerment finding, only that its explanation for making, or not making, such a finding must be based upon permissible statutory grounds – *i.e.*, the relationship between greenhouse gas emissions from new motor vehicles and public health or welfare.

Implications of the *Massachusetts* Decision

In the wake of the Supreme Court's decision, there has been both a call for EPA to take immediate action to begin regulating carbon dioxide emissions from motor vehicles and, perhaps more interestingly, intensified lobbying for Congressional action on climate change legislation. The former is hardly surprising. The Supreme Court held that carbon dioxide is an air pollutant, thereby setting the stage for EPA to initiate the regulatory process, or at least the process for deciding whether or not to make an endangerment finding. The latter, however, suggests that even advocates of regulation recognize that the victory of the decision may be a hollow one. If the goal is truly to reduce the atmospheric concentration of carbon dioxide and other greenhouse gases that scientists indicate are causing or contributing to global warming, and all of its attending effects, regulation under the Clean Air Act is not the answer. As discussed in greater detail below, the tools of the Clean Air Act are simply not well suited to address a global pollutant like carbon dioxide.

First, it is important to understand exactly what the Court's decision does, and does not, require.

- As noted above, the Court did not reach the issue of whether EPA must make an endangerment finding. On remand, however, if the Agency opts *not* to make an endangerment finding, it must articulate why there is such profound scientific uncertainty that it cannot make that finding.

- *If* the Agency does make an endangerment finding, it must then propose regulations to address greenhouse gas emissions from new motor vehicles. That is really the only true regulatory mandate of the Supreme Court's decision.
- Significantly, the Agency retains substantial discretion with respect to the content of any regulation. The Majority opinion states that "EPA has no doubt significant latitude as to the manner, timing, content and coordination of its regulations with those of other agencies." Slip op. at 30.
- The Supreme Court's decision does not address stationary sources and therefore does not require that EPA undertake any action with respect to the regulation of stationary sources.

EPA's Existing Statutory Authority to Regulate Air Pollutants

As noted above, the Court's decision could have far-reaching implications beyond simply the regulation of mobile sources under section 202 of the Clean Air Act. First, the Court's holding that greenhouse gases are "air pollutants" means that EPA has broad authority to regulate greenhouse gases under all the significant Clean Air Act programs, including the National Ambient Air Quality Standards (NAAQS), New Source Review (NSR), New Source Performance Standards (NSPS), Prevention of Significant Deterioration (PSD), stratospheric ozone (Title VI), and mobiles sources and fuels (Title II) programs. Second, the Court's constrained approach to the endangerment finding may limit, although not preclude, EPA's ability to decide not to regulate greenhouse gas emissions under those programs since they, like section 202, are triggered when the Administrator determines that an "air pollutant" causes or contributes to air pollution that "may reasonably be anticipated to endanger public health or welfare." Having the authority to regulate under existing law, however, does not mean that regulation will be effective.

National Ambient Air Quality Standards

Section 108 of the Clean Air Act requires the Administrator to publish and, "from time to time thereafter revise," a list of air pollutants: (1) emissions of which, in his judgment, cause or contribute to air pollution which may reasonably be anticipated to endanger public health or welfare; and (2) that are emitted from numerous or diverse mobile or stationary sources.[3] Once a pollutant is listed, EPA is required to establish primary and secondary air quality standards for that pollutant.

States deemed to be in attainment must develop State Implementation Plans ("SIPs") demonstrating how they will maintain compliance; nonattainment states must develop SIPs demonstrating how they will come into attainment with the standards "as soon as practicable" but no later than five years after designation.[4] States that fail to submit SIPs or to come into attainment within the statutory deadlines attain face potential sanctions, including the potential loss of highway funding, and a federal takeover of their CAA programs.

Although the argument could be made that CO_2 meets the statutory threshold for designation and regulation as a criteria pollutant, it is evident that this would make little sense from a regulatory perspective. If the standard were set at a level intended to force reductions in emissions, *i.e.*, at some atmospheric concentration below current levels (approximately 370-380 parts per million CO_2, then the entire country would be designated as being in nonattainment.[5] This would trigger the regulatory mechanisms of the NAAQS program – SIPs, NSR, reasonably available control technologies (RACT)to reduce emissions—but the reality is that none of the measures will have any effect in terms of bringing any individual State or county into attainment. Unless international emissions are also reduced, global CO_2 concentration will continue to increase and the entire United States would remain in nonattainment status. Even with international reductions, which are not currently occurring, the statutory deadline for compliance – a maximum of 12 years – is patently unrealistic and unachievable. This should be of concern to States that face potentially significant penalties for persistent nonattainment. For these reasons, it should be clear that the NAAQS program is ill suited to address a global pollutant like CO_2.

New Source Performance Standards

Section 111(b)(1)(a) of the Clean Air Act requires the Administrator to adopt new source performance standards for categories of emission sources that "cause[], or contribute[] significantly to, air pollution which may reasonably be anticipated to endanger public health or welfare." EPA is also required to review, and, if appropriate, revise, the NSPS every eight years to ensure that the standards continue to protect public health and the environment. CAA § 111(b). These standards are developed on a specific unit-by-unit basis, and apply to both attainment and nonattainment areas. Emission standards under the NSPS program must reflect "the degree of emission limitation achievable through the application of the best system of emission reduction" that has been "adequately demonstrated," while considering the "costs of achieving such reductions and any non-air quality health and environmental impact and energy requirements." CAA § 111(a)(2).

In the wake of the Supreme Court decision, EPA could issue sector-specific emissions standards for CO_2, assuming that it makes the necessary endangerment finding. These standards, however, by definition would not be economy-wide and furthermore would generally apply only to new sources. It is true that the Agency could, through multiple rulemakings, ultimately seek to regulate a number of industry sectors, but the process would certainly be a lengthy one extending over a period of many years. The standards themselves must be based upon the best demonstrated technology, which EPA has interpreted to mean technology that is in existence *and* widely commercially available. This could further limit the value of NSPS in terms of achieving significant and immediate reductions in emissions.

Cap and Trade Programs

Most proponents of regulation or legislation to address greenhouse gas emissions argue that the most effective means of achieving reductions is through a market- based cap and trading program. In a more limited context, EPA has successfully implemented a trading program to reduce sulfur dioxide (SO2) from utilities under the Acid Rain program specifically authorized by Title IV of the Clean Air Act. It subsequently developed a cap and trade program for nitrogen oxides (NOx) under the NOx SIP call using its authority under section 110 of the Act. More recently, the Agency promulgated the Clean Air Interstate Rule (CAIR) pursuant to its authority under section 110, to further reduce NOx and SO2 from power plants. These programs have generally been upheld by the courts or not challenged.

Whether or not EPA has the authority to develop a cap and trade program for greenhouse gases, however, may still be at issue. Experience with the NOx SIP call and CAIR suggest that a trading program under section 110 of the Act would likely survive judicial challenge. That would first require the listing and regulation of CO_2 as a criteria pollutant, which as discussed above, makes little sense. Alternatively, EPA could use its authority under the NSPS provisions of section 111 of the Act to create a cap and trade program, as it did recently for mercury in the Clean Air Mercury Rule. However, the mercury rule, and specifically EPA's assertion of authority under section 111 to create a cap and trade program rather than unit-specific standards, is being challenged in the D.C. Circuit. Until that fundamental legal question is resolved, EPA's ability to craft an effective cap and trade program under existing law remains unclear.

CONCLUSION

While the *Massachusetts v. EPA* decision put to rest the question of whether greenhouse gases are "air pollutants" under the Clean Air Act, this in and of itself, will do little to address climate change in a meaningful way. The Clean Air Act's existing regulatory tools were simply not designed to address global pollution. Climate change is an international problem; it demands an international solution. It is a national policy issue; it demands a national policy solution.

End Notes

[1] A substantial portion of the Majority's opinion focuses on the issue of standing and, in particular, whether the petitioners in this case have satisfied the elements of Article III standing under the Constitution. After setting forth a novel theory of standing premised upon "a special solicitude" for the State of Massachusetts based upon its "stake in protecting its quasi-sovereign interests," the Stevens Majority concludes essentially that loss of Massachusetts coastline constitutes sufficient injury in fact that might be traced, in some small part, to climate change and redressed, again in some small part, by future regulation of emissions from new motor vehicles. Chief Justice Roberts, in a dissenting opinion joined by Justices Scalia, Thomas and Alito, would have rejected the challenge to EPA's action as nonjusticiable. The dissent notes that there is no basis in law for the Majority's "special solicitude" for the State of Massachusetts in its standing analysis. Furthermore, as the dissent sets forth in some detail, the State's injury is neither particularized, nor imminent; the injury cannot reasonably be traced to the lack of regulation of greenhouse gas emissions from new motor vehicles, particularly given the numerous and complex factors that affect all predictions with respect to climate change; and, finally, the injury cannot be meaningfully addressed by the action sought – regulation of new motor vehicles – because emissions from new motor vehicles account for only a minute percentage of the global atmospheric concentration of carbon dioxide. For these reasons, the State of Massachusetts and the other petitioners cannot meet the three requirements of Article III standing.

[2] As Justice Scalia noted in footnote 2 of his dissenting opinion, this interpretation of the language of the definition of "air pollutant" would make little sense as it would then follow that "*everything* airborne, from Frisbees to flatulence, qualifies as an air pollutant." Scalia dissent at 10.

[3] The six listed criteria pollutants are: ozone, nitrogen dioxide, particulate matter, sulfur dioxide, carbon monoxide and lead.

[4] The statute provides that States must come into attainment within five years, but it authorizes EPA to grant one five-year extension. Additionally, states can seek two additional one-year extensions. Thus, under the CAA, can get up to twelve years to attain the (non-ozone) NAAQS.

[5] Alternatively, if the standard were set above current levels of CO_2, the entire country would, at least for the short term, be classified as an attainment area and no regulatory mechanisms to reduce emissions would be triggered. This result, would be short-lived, however, as emissions from China and India continue to increase dramatically. Thus, regardless of what individual States or counties do with respect to their CO_2 emissions, global atmospheric concentrations will continue to increase.

In: Climate Change Litigation and Law
Editors: Jean-François Masson ISBN: 978-1-60876-089-3

Chapter 8

HEARING ON IMPLICATIONS OF SUPREME COURT'S EPA DECISION- REILLY TESTIMONY

William K. Reilly
Senate Committee on Environment and Public Works

Madame Chairman, Senator Inhofe, Members of the Committee, my name is William K. Reilly. I served as Administrator of the U.S. Environmental Protection Agency under President George H. W. Bush, from 1989 to early 1993.

Thank you for the opportunity to appear before the Committee. I applaud your initiative on this urgent and compelling matter. And I am pleased to appear with my distinguished successors, Administrators Johnson and Browner. With your permission, I will submit my formal statement for the record.

Though I am appearing on my own behalf, I note for the record that since 2002 I have co-chaired the bipartisan National Commission on Energy Policy. Our 2004 report recommended a mandatory program to reduce greenhouse gases with various safeguards, as well as addressing many other issues in energy policy, including oil security, supply, efficiency, technology, and more. The Commission's staff continues to confer widely with Members of the Senate and the House on these matters. Were I the EPA Administrator, or a Member of Congress, I would recognize the extensive research and inclusive membership of the Commission, and take the Commission's recommended policy on climate change as both an effective national starting point and also as the policy proposal that stands the most realistic chance of being enacted. Extensive, detailed research financed generously by the Hewlett and other foundations underlies the Commission's recommendations. The Energy Information Agency has analyzed the costs of the Commission's proposals and concluded they are reassuringly modest. So my advice to the Congress and the Administration is, take a hard look at the Commission's report.

You've asked me to discuss EPA's role in the wake of the Supreme Court decision holding that EPA has authority as a matter of law to regulate carbon dioxide. I'm not going to delve into the legal reasoning or the language of the Court's decision. I read it as expansive with regard to taking action on harmful pollutants. Suffice to say, the law has now been settled and EPA does have the authority. I might add that if I were EPA Administrator, I would welcome that authority.

The Court's decision is of immense consequence and signals the growing significance of concern about climate change. The decision represents the intersection of science and public policy. All that follows must be grounded in good science. Indeed, the science is becoming increasingly compelling. This Administration, as well as those of President Clinton and of President Bush, whom I served, deserves great credit for their support for the scientific research underpinning our understanding of climate change. The nation has spent billions of dollars to get to this point. This year's reports of the Intergovernmental Panel on Climate Change (IPCC) affirm the high degree of confidence that hundreds of participating scientists have in the scientific findings.

When I was named EPA Administrator, one of my first briefings was on climate change, by Dr. Frank Press, then president of the National Academy of Sciences. EPA also had underway in the policy office a couple of reports on the effects of climate change and policy options to address them. Most of this work and the work of others was premised on computer modeling and projections, and the findings were subsequently subjected to a lively debate about the assumptions inherent in the models and their accuracy.

We are no longer limited to relying on computer models. As the IPCC reports made clear, we are already seeing signs of climate change and variability associated with the buildup of greenhouse gases in the atmosphere. The models have been greatly refined and it is my understanding that they now comport well with the mounting evidence from field observations and related research in any number of areas, from wildlife behavior to snow pack and melting glaciers, to sea level rise, changes in precipitation, temperature records that cannot be dismissed as merely the result of urban heat island effect, and more.

Not all matters are resolved, of course. Questions remain about the timing, the magnitude, and the local impact of the effects, and there is still much to learn about how the systems function to shape climate on earth. But given what the IPCC reported, we cannot afford to wait until all matters are resolved. That was the thrust of the *amicus brief* that I submitted in concert with Administrators Browner, Costle and Train. We have not required in the past, nor should we require in the future, an unrealistic level of certainty in addressing serious and urgent problems such as climate change, even as we acknowledge that we may have to change course, to take more or less aggressive action as further information becomes available. To delay action on climate change means that down the road, what we do will necessitate more expensive and more draconian measures.

In light of this evidence and the Supreme Court's decision, what should EPA do?

The Court's decision confronts the EPA with a choice of contesting the scientific consensus regarding the causes of global warming, which it has conceded, and then of asserting or rejecting in its judgment the merits of regulating what the Court has determined to be a "pollutant." It is difficult to see how the Agency can now refrain from moving forward to regulate greenhouse gases from automobiles and by implication from other sources as well. The practical realities must be faced, however. The regulation of greenhouse gases is hugely consequential for many sectors of the economy, as for the health and well-being of Americans and others. To ask EPA to assume the full burden of recommending in a regulatory program the full gamut of measures necessary to the task is unrealistic. It is particularly so given that the Agency is part of an Administration that has consistently declined to embrace the regulation of carbon dioxide. One cannot expect a robust rulemaking in such a circumstance. The situation cries out for Congressional action and that, in my view, is a principal merit in the Court's decision.

So it is enormously ambitious to expect that a regulatory agency alone, even one as well-versed as U.S. EPA, can craft a regulatory regime governing something so far-reaching with such substantial impacts on our economy and industry, on the natural resources on which we depend, on U.S. foreign policy and the prospects for development in the world's poorest countries. And yet that is the challenge.

I would note that regarding the Clean Air Act of 1990, with which I had something to do, it took more than a decade for this legislation to come together—on acid rain, standards for air toxics, upper atmospheric ozone depletion, and the other issues it addressed—and for the political context to ripen. EPA staff had spent the 1980s preparing the analyses which they knew would one day be needed when the moment came that clean air legislation stood a serious chance of passage. Between my swearing in and the President's submission of a comprehensive legislative proposal to Congress, we required just four months.

That we could move so quickly is a tribute to the substantial and rigorous work done by the Agency during the 1980s, including seminal work on emissions trading with Environmental Defense Fund and Resources For the Future, analyses of costs and benefits, and more. The acid rain trading program, which emerged from the 1990 clean air law and which by all accounts has been a resounding success, is the reference case for our way into a cap-and-trade regime for carbon dioxide.

I would be remiss if I didn't state my high regard for the senior career staff in the EPA's Air Office and those who served in what was our Office of Policy, Planning and Evaluation. Contrary to the belief in some quarters, they are not eco-cowboys who find something to regulate under every rock they lift. They are smart, creative, experienced, and dedicated people, and they grasp full well the implications and tradeoffs, the costs and benefits associated with fulfilling their mission to clean up and safeguard our nation's air. The country has been well served by these civil servants, and I expect no less from them in dealing with climate change.

During the past few years they have been carrying out the research and analysis of options for regulating greenhouse gases just as in the 1980s they prepared analysis of directions a new clean air act might take. What are realistic targets and timetables? Would analysis show a carbon tax to be more effective? Or a cap-and-trade system? What are the downsides? Clearly, a carbon tax is beyond EPA's jurisdiction and I would be wary of recommending one if the implication was that EPA would therefore take no steps to regulate carbon pursuant to the Agency's authority.

For an emissions trading program, do we want a safety valve to contain costs? What is the point of regulation, which sectors? How would permits be distributed, how many, and based on what criteria? What are the implications of these approaches? What is the state of technology, the connections with other emissions of concern, notably mercury, sulfur dioxide and other criteria pollutants? EPA staff are more prepared than we know to put forth the options for designing a carbon policy. They have spent more than 15 years preparing for this moment. So my advice is, challenge them to present the policy options.

I do not expect that even with heroic efforts, these matters will translate immediately into a regulatory program. There is a lengthy regulatory process, as you know, involving not just interagency reviews, but consultations with states and industrial sectors and others outside the federal family. There are formal administrative procedures to follow and a record to prepare, and that could be substantial for an issue as complex as regulating carbon dioxide. And of course, there is the potential for litigation once a rule is adopted. Bill Ruckelshaus once observed that 4 of every 5 major EPA decisions wind up in court.

To be sure, there is much activity in Congress, both in the Senate and the House, and I believe that ultimately the issue of climate change needs to be addressed by Congress. That said, there is no reason for EPA to delay. On a parallel track, EPA should begin the regulatory process for carbon dioxide. This would be a timely and useful step, and would both inform the legislative debate and keep pressure on Congress to continue its work. At the same time, EPA's efforts now will prepare the Agency for quick progress in implementing any legislation after enactment, as was the case after passage of the 1990 Clean Air Act.

I would welcome the full involvement of the President and the Administration in these deliberations. Indeed, that would help engage some who are still skeptical about the science or the nation's ability to take the issue head on, and would help ensure that economic impacts, foreign policy concerns, and other important considerations are taken into account. I do not support the case for awaiting the arrival of a new President and a new Administration to address this issue.

Besides beginning this process, there are a number of other important steps the Agency should take.

First, California has a request pending for a waiver to reduce CO2 emissions from automobile fuels by 30 per cent, beginning with the 2016 model year. I understand that process is getting underway, and I would urge all due speed. California's proposal is the product of a bipartisan effort and has tremendous public support. The Supreme Court's decision should remove any roadblocks with respect to the review process.

Second, I would urge EPA to take a good look at what Governor Schwarzenegger of California has called for, via Executive Order, to set a low carbon standard for fuels. This seems to me a very innovative approach to ensure that, as we struggle with the very real issue of oil security, we do not substitute for what we now use new fuels with worse greenhouse gas impacts. I doubt any regulatory entity has the experience with fuels that EPA does, with a world-class mobile source laboratory in Michigan, and from the Agency's prior experience in removing lead from gasoline, and the work regarding particulates, ozone, and the recent well-regarded rule lowering sulfur content in diesel fuel. I applaud Administrator Johnson's decision and the support he had from the Administration in getting this rule out. It is one of the most significant contributions to clean air. I should point out that the low carbon standard for fuels initiative also enjoys widespread support in California, including that of ConocoPhillips, the nation's largest refiner.

Third, I would like to see EPA develop the regulatory approach for carbon capture and sequestration. That is the key to using our abundant coal resources and to ensuring that other countries with substantial coal reserves do not undo all that we might accomplish in reducing greenhouse gases. Because of its experience and its record in dealing with underground injection, EPA's Water Office in partnership with other parts of the Agency, and most importantly with the Department of Energy, is well-suited to undertake this task. It is my understanding, however, that although the Energy Department has substantial funding to develop this critical technology, EPA has little, making it difficult to draw on the Agency's experience and credibility with the various stakeholders. I would add that many in the power industry want to see a regulatory program for carbon capture and sequestration quickly, lest the absence of a regulatory framework delay testing and deployment of this promising, indeed, *essential* technology. Not just America, but China, India and other coal-rich nations stand in urgent need of carbon sequestration technology.

Fourth, a number of states are taking action on greenhouse gas reductions and I would ensure that EPA is well-versed on these actions and the regional compacts that are beginning to emerge. A national program invariably invites the question of federal pre-emption and that will surely surface with respect to regulating carbon dioxide. Moreover, it behooves us to learn from what the states are doing. I would add that virtually every law in the EPA administrator's portfolio had origins at the state level, none more so than California and air quality.

I would also call your attention to the good work of the Center for Climate Strategy, which has been working with a couple of dozen states to prepare greenhouse gas inventories, consider policy options, costs, and associated measures, with an eye toward state action and the role of states in implementing a national program.

Fifth, I would urge EPA to become fully versed in the European Union's emissions trading program. There is evidence that too many credits were distributed in the first round of permit allocations, resulting in less than optimal performance, a drop in permit values, and a windfall for some firms. We need to learn from that experience in this area of allocation, lest we repeat it.

Sixth, I would encourage EPA and others in the federal government to remain on top of climate developments in China and India, two critical countries with respect to greenhouse gas emissions. My experiences in China with the Energy Foundation's China Sustainable Energy Program suggest that although not now party to any international protocols requiring it to reduce greenhouse gases, China is well aware of the potential impacts and is taking measures to improve efficiency of energy use. We will need to engage these countries in international forums and we would be well-served by following developments in those countries closely, and by establishing contacts at the technical level which I believe the Chinese would welcome.

In closing, let me state that as important as a mandatory national program is to reduce carbon dioxide emissions, it is but one measure we need. If scientists are right about the impact of doubling carbon dioxide in the atmosphere over pre-industrial levels, which is where we are heading under business as usual—indeed, we may see a tripling or more if we don't take action soon—then we will be called on to make far more drastic cuts in greenhouse gas emissions, well below today's level, even while we continue to grow in population and economic activity.

That goal would be achievable only with a suite of policies and programs going beyond a cap-and-trade system. We will need substantially improved mileage standards for automobiles, trucks, and other vehicles, which will help on oil security as well. We will want a national renewable portfolio standard to advance deployment of renewable energy technologies, much as a couple of dozen states have already enacted. We will need to invest heavily in technology research, development, and deployment. I mentioned carbon capture and sequestration. Cellulosic ethanol and other promising bio-fuels also merit increased funding. We will want to move aggressively on efficiency standards. Some 22 or so are currently under development at the Department of Energy. We will need to involve the states, for they have a major role in building codes, water resource management, land use and transportation planning. They build and operate public buildings and institutions, and we now know that for all of the design techniques to improve energy efficiency, most of the savings come in operations over the life of a facility.

And as important as mitigation is in fending off the worst scenarios, we will need to prepare to adapt, for the science is telling us that we are seeing the effects today and we know that carbon dioxide and other greenhouse gases are long-lived in the atmosphere, so more elevated concentrations are already built into the system.

Congress has engaged the climate issue in a direct and serious way. Within the next several months, there may be a window of opportunity for legislation on climate change. You know better than I. After that, we may well see 2008 campaign politics adding to the hurdles. That would make EPA's endeavors all the more important. I wish you and your colleagues success. The country, indeed the entire world, is counting on it.

Thank you.

In: Climate Change Litigation and Law
Editors: Jean-François Masson
ISBN: 978-1-60876-089-3

Chapter 9

THE SUPREME COURT'S CLIMATE CHANGE DECISION: *MASSACHUSETTS V. EPA*

Robert Meltz
American Law Division

SUMMARY

On April 2, 2007, the Supreme Court handed down *Massachusetts v. EPA*, its first pronouncement on climate change. By 5-4, the Court held that (1) Massachusetts had standing to sue, (2) Section 202 of the Clean Air Act authorizes EPA to regulate emissions from new motor vehicles on the basis of their possible climate change impacts, and (3) Section 202 does *not* authorize EPA to inject policy considerations into its decision whether to so regulate. The Court's decision leaves EPA with three options under the section: find that motor vehicle greenhouse gas emissions may "endanger public health or welfare" and issue emission standards, find that they do not satisfy that prerequisite, or decide that climate change science is so uncertain as to preclude making a finding either way. The decision also has implications for other climate-changerelated litigation, particularly a pending suit seeking to compel EPA regulation of greenhouse gas emissions from stationary sources of emissions.

On April 2, 2007, the Supreme Court handed down *Massachusetts v. EPA*, its first pronouncement on climate change.[1] By a narrow 5-4 margin, the Court held three things: that (1) Massachusetts had standing to sue, (2) the Clean Air Act (CAA) authorizes EPA to regulate emissions from new motor vehicles on the basis of their climate change impacts, and (3) the act does *not* authorize EPA to inject policy considerations into its decision whether to so regulate. The decision does not compel EPA to regulate greenhouse gas (GHG) emissions from new motor vehicles, but it does limit the range of options available to the agency that would justify not doing so.

This report traces the events leading up to the Court's decision, then describes what the decision says. It then lays out the decision's implications both within the CAA mobile-source program and elsewhere.

EPA's Denial of the Section 202 Petition

In 1999, 19 organizations petitioned EPA to regulate emissions of GHGs (CO_2, methane, nitrous oxide, and hydrofluorocarbons) from new motor vehicles. The petition cited the agency's alleged mandatory duty to do so under CAA Section 202(a)(1).[2] That provision directs the EPA Administrator to prescribe by regulation

> standards applicable to the emission of any air pollutant from any class or classes of new motor vehicles ... which, in his judgment cause, or contribute to, air pollution which may reasonably be anticipated to endanger public health or welfare.

In 2003, after receiving about 50,000 comments, EPA denied the Section 202 petition.[3] Much of the agency's rationale followed a General Counsel memorandum issued the same day.[4] Contrary to its Clinton Administration precursor,[5] this General Counsel memorandum concluded that the CAA does *not* grant EPA authority to regulate CO_2 and other GHG emissions based on their climate change impacts.

Massachusetts v. EPA in the D.C. Circuit

EPA's denial of the Section 202 petition prompted a suit, *Massachusetts v. EPA*, in the D.C. Circuit. Petitioners were twelve states (CA, CT, IL, MA, ME, NJ, NM, NY, OR, RI, VT, WA); three cities (New York, Baltimore, and Washington, D.C.); two U.S. territories (American Samoa and Northern

Mariana Islands); and several environmental groups. Opposing the challenge, besides EPA, were 10 state intervenors (AK, ID, KS, MI, ND, NE, OH, SD, TX, UT), plus several automobile- and truck-related trade groups.

In 2005, a split panel rejected the suit.[6] The two judges supporting rejection, however, did so for different reasons. Judge Randolph concluded that EPA had properly exercised its discretion in choosing not to wield its Section 202 authority. As to this discretion issue, recall that CAA Section 202(a)(1) directs the EPA Administrator to prescribe standards for any motor vehicle emissions that "in his judgment" cause harmful air pollution. Judge Randolph read "in his judgment" broadly to allow EPA consideration of not only scientific uncertainty about the effects of GHGs but also policy considerations that justify not regulating. Thus, EPA in his view was entitled to rely, as it did, on such factors as the current Bush Administration's policy preference for voluntaryGHG control measures, and its belief that regulating motor vehicle emissions was a piecemeal, hence inefficient, approach to dealing with climate change. By contrast, Judge Sentelle, the other judge supporting rejection of the petition, simply held that petitioners lacked standing.[7]

In dissent, Judge Tatel asserted that Massachusetts had demonstrated standing through past and future loss of shore land as a result of climate-change-induced sea level rise. On the merits, he found that EPA has authority under Section 202(a)(1) to regulate GHG emissions.[8] He further concluded that EPA's 202(a)(1) discretion does not extend to policy considerations, as Judge Randolph held, but relates exclusively to whether the emissions cause harmful air pollution.

Massachusetts v. EPA in the Supreme Court

It was somewhat of a surprise that the Supreme Court agreed to review the D.C. Circuit decision in *Massachusetts v. EPA*. There was no split in the circuits, which often disposes the Court not to take a case, and the D.C. Circuit majority had not even ruled on the key issue: whether Section 202(a)(1) authorizes regulation of GHG emissions. Moreover, grants of certiorari over the opposition of the United States, as here, are rare. But as the Supreme Court stated in its decision, "the unusual importance of the underlying issue persuaded us to grant the writ."

The Court ruled 5-4 for petitioner states and environmental groups on all three issues in the case: standing, authority (whether "air pollutant" includes GHG emissions), and discretion (whether "in his judgment" allows policy considerations).[9] Justice Kennedy provided the fifth vote by joining Justice Stevens' opinion for the Court's "liberal/moderate" bloc. The ruling in favor of petitioners was forecast early in the majority opinion by its opening sentences: "A well-documented rise in global temperatures has coincided with a significant increase in the concentration of carbon dioxide in the atmosphere. Respected scientists believe the two trends are related." (Nor did the dissenters dispute this.)

Most of the decision is devoted to standing, an issue that is ubiquitous in climate change litigation. At the outset, the Court found that petitioners had two factors in their favor. First, the CAA specifically authorizes challenges to agency action unlawfully withheld, such as this one.[10] A litigant to whom Congress has accorded such a procedural right, said the Court, can assert that right without meeting the normal standards for standing. Second, the Court found it "of considerable relevance" that the petitioner injury on which it focused — Massachusetts's loss of shore land from global-warming-induced sea level rise — was that of a sovereign state rather than a private entity. States are "not normal litigants for the purposes of invoking federal jurisdiction," said the Court, noting their quasi-sovereign duty to preserve their territory.

Having described petitioners' favored position with regard to standing, it was curious that the Court then undertook a fairly traditional standing analysis. As to the first prong of the black-letter standing test — whether plaintiff has demonstrated actual or imminent "injury in fact" of a concrete and particularized nature — the Court homed in on Massachusetts's status as owner of much of the commonwealth's shore land. That this injury may be widely shared with other coastal states does not disqualify this injury, said the Court; it is nonetheless concrete.

The second prong of the standing test is causation, requiring that the injury of which the plaintiff complains is fairly traceable to the defendant. EPA did not dispute the existence of a causal relationship between GHG emissions and climate change. It did argue, however, that any reduction in GHG emissions achieved through the current litigation would be too tiny a fraction of worldwide GHG emissions to make a cognizable difference in climate change. In an important ruling that may be of benefit to environmental plaintiffs in many contexts, the Court held that even an agency's refusal to take a "small incremental step" that would result in only a modest reduction in worldwide GHG emissions, is enough for standing purposes.

The third and final prong of the standing test is redressability, demanding that the remedy sought by the plaintiff is one that is likely to redress his injury. In this case, the remedy sought is EPA regulation of GHG emissions from new motor vehicles. The Court found that this remedy satisfied redressability because while it would not by itself reverse climate change, it would nonetheless slow or reduce it. Nor, given the "enormity" of the potential effects of climate change, was it relevant to the Court that the full effectiveness of the remedy would be delayed until existing cars and trucks on the road were largely replaced by new ones.

Given the large number of pages devoted by the majority opinion to standing, its discussion of the two CAA issues in the case seems strangely brief. On the authority question, the CAA's "sweeping" definition of "air pollutant" — embracing "*any* air pollutant ... including *any* physical, chemical ... substance or matter which is emitted into or otherwise enters the ambientair"[11] — simply could not be squared, in the Court's view, with EPA's position that GHGs are not included. The Court rejected EPA's argument that federal laws enacted following enactment of this statutory language — laws emphasizing interagency collaboration and research — suggest that Congress meant to curtail EPA's power to use mandatory regulations in addressing air pollutants. Nor was the Court impressed with EPA's contention that "air pollutant" in the CAA could not include vehicle GHG emissions because EPA standards for such emissions could be satisfied only by improving fuel economy, a job EPA asserted was assigned solely to the Department of Transportation under a different statute (the Energy Policy and Conservation Act[12]).

Finally, on the discretion issue,themajority concluded that "in his judgment" allows the EPA Administrator to consider only whether an air pollutant "may reasonably be anticipated to endanger public health or welfare," not the agency's policy preferences. Policy considerations, at least those that led EPA to reject the petition, "have nothing to do with whether greenhouse gas emissions contribute to climate change." Thus, said the Court, EPA can avoid taking further action in response to the Section 202 petition "only if it determines that greenhouse gases do not contribute to climate change or if it provides some reasonable explanation as to why it cannot or will not exercise its discretion." In sum, whether EPA decides to act or not, "[it] must ground its reasons for action or inaction in the statute."

Accordingly, the Court reversed the D.C. Circuit opinion, and remanded the case to that court for further proceedings. It is almost certain that the D.C. Circuit will issue an order sending the petition back to EPA.

A four-justice dissent by Chief Justice Roberts vigorously disputed the majority's finding of standing. A four-justice dissent by Justice Scalia disputed that "air pollutant" in Section 202 includes GHGs.

Implications of the Supreme Court Decision for Mobile Sources

The Court's decision leaves EPA with three options under Section 202: make a finding that motor vehicle GHG emissions may "endanger public health or welfare" and issue emissions standards, make a finding that they do not satisfy that prerequisite, or decide that the science is so uncertain as to preclude making a finding either way (or cite some other "reasonable explanation" why it will not exercise its discretion either way).

The EPA Administrator has said following *Massachusetts v. EPA* that while the decision bars EPA use of policy considerations as a basis for *denying* the petition, it left open whether the agency can invoke them later when actually *writing* the regulations, should it make an endangerment finding.[13] In this regard, it should be noted that CAA Section 202 does not explicitly impose any stringency or other criteria on GHG emission standards promulgated under the section. While Section 202 states numerical formulae for emissions of carbon monoxide, hydrocarbons, and nitrogen oxides, none is stated for GHG emissions other than nitrogen oxides.

Reflecting the apparently wide latitude EPA has in setting mobile-source GHG emission standards under Section 202 (other than for nitrogen oxides), commentators have suggested that EPA, following an endangerment finding, could be creative. The agency might, some have opined, set (1) voluntary standards, (2) standards pegged to the CAFÉ standards for fuel economy, or (3) standards that must be complied with only after the President certifies that developing nations have put adequate GHG emission limits into effect, among other options. It is not clear whether courts would uphold all of these innovative interpretations. In any event, on May 14, 2007, the President asked the EPA Administrator, working with the Departmentsof Transportation, Energy, and Agriculture, to have CAA regulations limiting vehicle GHG emissions in place by the end of 2008, and to use the President's 2007 State of the Union proposal for raising the CAFÉ standards as a guide.[14] This appears to mean that EPA will be making the prerequisite finding that motor vehicle GHG emissions endanger public health or welfare.

Other Implications of the Supreme Court Decision

The Court's ruling in *Massachusetts v. EPA* resonates well beyond its four corners. At this early date, however, only broad predictions are possible. Most obviously, the finding of standing likely will be pivotal to the fortunes of plaintiffs in other climate change litigation. A key question will be whether the finding was contingent, as could be argued, on the existence of astatepetitioner and thepresence in the CAA of an explicit provision allowing the filing of administrative petitions. In cases seeking injunctive relief, this query is relevant only for private plaintiffs not joined by state plaintiffs. In injunctive cases with mixed private and state plaintiffs, a court can focus on the latter; as *Massachusetts v. EPA* instructs, it is only necessary that one plaintiff in such a case have standing. The decision also helps rebut other no-standing arguments often made in climate change litigation (e.g., that even were the defendants to reduce their GHG emissions substantially, it would make but an insignificant contribution to mitigating climate change).

Elsewhere, the Court's ruling upholding CAA coverage of GHG emissions from mobile sources improves the prospects of litigation seeking to have EPA restrict GHG emissions from *stationary* sources, such as coal-fired power plants and factories.[15] The stationary-source provisions of the CAA use terms similar to that of Section 202 — in particular, "air pollutant," "in his judgment," and "may reasonably be anticipated to endanger public health and welfare."[16] Contrariwise, the ruling could undermine pending cases making federal common law claims against entities alleged to contribute to climate change, such as coal-fired power plants and auto manufacturers, since it strengthens the argument that Congress in the CAA intended to leave no room for courts to develop overlapping federal common law restricting GHG emissions. Ironically, this means that the victory for the "environmental" side in *Massachusetts v. EPA* enhances the possibility of a defeat for that side in the federal common law cases.

End Notes

[1] Recent years have seen an explosion of litigation involving climate change, *Massachusetts v. EPA* being just one (though the most important) example. For a review of all the significant climate change litigation in the U.S., decided and pending, see CRS Report RL32764, *Climate Change Litigation: A Growing Phenomenon*, by Robert Meltz.

[2] 42 U.S.C. § 7521(a)(1).

[3] EPA, Control of Emissions from New Highway Vehicles and Engines, 68 Fed. Reg. 52922 (September 8, 2003).

[4] Memorandum from Robert E. Fabricant, EPA General Counsel, to Marianne L. Horinko, EPA Acting Administrator, EPA's Authority to Impose Mandatory Controls to Address Global Climate Change Under the Clean Air Act (August 28, 2003).

[5] Memorandum from Jonathan Z. Cannon, EPA General Counsel, to Carol M. Browner, EPA Administrator, EPA's Authority to Regulate Pollutants Emitted by Electric Power Generation Sources (April 10, 1998).

[6] 415 F.3d 50 (D.C. Cir. 2005).

[7] The test for whether a plaintiff in federal court has standing is described in greater detail in the following discussion of the Supreme Court's decision on appeal of the D.C. Circuit ruling.

[8] 415 F.3d at 62 (emphasis added by court).

[9] 127 S. Ct. 1438 (2007).

[10] CAA § 307(b)(1), 42 U.S.C. § 7607(b)(1).

[11] Emphasis added by the Court.

[12] 49 U.S.C. § 32902.

[13] The EPA Administrator is apparently referring to the Court's statement that "We need not and do not reach the question ... whether policy concerns can inform EPA's actions in the event that it makes [an endangerment finding]."

[14] *Briefing by Conference Call on the President's Announcement on CAFÉ and Alternative Fuel Standards*, May 14, 2007 (statement of EPA Administrator Stephen Johnson), available at [http://www.whitehouse.gov/news/releases/2007].

[15] New York v. EPA, No. 06-1322 (D.C. Cir. Sept. 13, 2006) (severed from pre-existing case by order of the court).

[16] See, e.g., CAA § 108(a)(1)-(2), 42 U.S.C. § 7408(a)(1)-(2) (requiring the EPA Administrator to maintain a list of each "air pollutant" "emissions of which, in his judgment, cause or contribute to air pollution which may reasonably be anticipated to endanger public health or welfare," and then issue air quality criteria and national ambient air quality standards for that air pollutant).

CHAPTER SOURCES

The following chapters have been previously published:

Chapter 1or2 – This is an edited, excerpted and augmented edition of a United States Congressional Research Service publication, Report Order Code RL34099, dated February 10, 2009.

Chapter 2or3 – This is an edited, excerpted and augmented edition of a United States Congressional Research Service publication, Report Order Code RL31931, dated December 8, 2008.

Chapter 1or3 – This is an edited, excerpted and augmented edition of a United States Congressional Research Service publication, Report Order Code RL32764, dated April 7, 2008.

Chapter 4 - These remarks were delivered as Statement of Carol M. Browner, before the Committee on Environment and Public Works, U.S. Senate, dated April 24, 2007.

Chapter 5 – These remarks were delivered as Statement of David Doniger, Policy Director and Senior Attorney, Climate Center, Natural Resources Defense Council, before the Committee on Environment and Public Works, U.S. Senate, dated April 24, 2007.

Chapter 6 – These remarks were delivered as Statement of Peter Glaser, partner in the Washington D.C., office of Troutman Sanders LLP., before the Committee on Environment and Public Works, U.S. Senate, dated April 24, 2007.

Chapter 7 – These remarks were delivered as Statement of Ann R. Klee Crowell & Moring, before the Committee on Environment and Public Works, U.S. Senate, dated April 24, 2007.

Chapter 8 – These remarks were delivered as Statement of Honorable William K. Reilly, before the Committee on Environment and Public Works, U.S. Senate, dated April 24, 2007.

Chapter 9 - This is an edited, excerpted and augmented edition of a United States Congressional Research Service publication, Report Order Code RS22665, dated May 18, 2007.

INDEX

A

B

C

D

E

F

J

K

L

M

N

O

P

Q

R

S

T

U

V

W

Y